120 SOLVED SURVEYING PROBLEMS

for the California Special Civil Engineer Examination

Peter R Boniface, PhD, PLS

The Power to Pass™
www.ppi2pass.com

Professional Publications, Inc. • Belmont, California

Benefit by Registering This Book with PPI

- Get book updates and corrections.
- Hear the latest exam news.
- Obtain exclusive exam tips and strategies.
- Receive special discounts.

Register your book at **www.ppi2pass.com/register**.

Report Errors and View Corrections for This Book

PPI is grateful to every reader who notifies us of a possible error. Your feedback allows us to improve the quality and accuracy of our products. You can report errata and view corrections at **www.ppi2pass.com/errata**.

Cover drawings adapted with permission from *Staking Manual for Surveying*, Pfeiler and Associates Engineers, Inc.

120 SOLVED SURVEYING PROBLEMS FOR THE CALIFORNIA SPECIAL CIVIL ENGINEER EXAMINATION

Current printing of this edition: 5

Printing History

edition number	printing number	update
1	3	Minor corrections.
1	4	Minor corrections. Copyright update.
1	5	Minor corrections.

PPI
1250 Fifth Avenue, Belmont, CA 94002
(650) 593-9119
www.ppi2pass.com

Library of Congress Cataloging-in-Publication Data
Boniface, Peter R., 1937–
    120 solved surveying problems for the California special civil engineer examination /
    Peter R. Boniface.
        p.   cm.
    ISBN: 978-1-59126-016-5
    1. Surveying--California--Examinations, questions, etc. 2.
    Surveyors--Certification--California-- Study guides. I. Title: One hundred twenty solved
    surveying problems for the California special civil engineer examination. II. Title.

TA537.B64   2005
526.9′076--dc22
                                                                    2004057314

Table of Contents

Preface and Acknowledgments

The problems in *120 Solved Surveying Problems for the California Special Civil Engineer Examination* are based on the Engineering Surveying Test Plan as published on the website of the California Board for Professional Engineers, Land Surveyors, and Geologists. The exam syllabus is extremely broad and touches on topics that most civil engineers have not studied during their undergraduate education. You should therefore take every opportunity to prepare and practice!

The 120 problems in this book mirror the actual exam problems in subject matter, length, and degree of difficulty. Although past examinations are not generally made publicly available, this book does cover all topics that appear on the exam—equipment and field activities, field measurements, calculations, and office procedures. The problem types presented here approximate the percentages given by the board-adopted test plan.

I would like to thank Reza Mahallati, PE, for performing the technical review of this book. I would also like to thank the PPI production and editorial staffs—especially Heather Kinser (editor), Miriam Hanes (typesetter), Amy Schwertman (illustrator), and Matt Breault (calculation checker)—for their hard work and patience in helping to put this book together.

Peter R. Boniface, PhD, PLS

Introduction

Preparing for the Exam

120 Solved Surveying Problems is designed to prepare you for the California Special Civil Engineer Examination in surveying. California is currently the only state that requires candidates for licensure as civil engineers to pass such an exam in addition to other registration requirements. Whether you enter into a program of self-study or of formal instruction, as suits your personal learning approach, there is no better way to increase your proficiency and improve your chances of passing this state exam than by solving practice problems.

Begin your exam preparation by reviewing a wide range of textbooks in this field. You'll need a solid understanding of the fundamentals and principles of engineering surveying. As others have found, excellent results come from studying *Surveying Principles for Civil Engineers* and *1001 Solved Surveying Fundamentals Problems*, both produced by PPI. *120 Solved Surveying Problems* is designed to complement those references.

Next, make intensive problem-solving practice a key to your success. To bolster your preparedness, use *120 Solved Surveying Problems* to practice until you feel confident. Each problem in this book is fully worked-out and details the relevant points and essential steps for a successful solution. You can then simulate the exam experience using *Civil Surveying Sample Exams for the California Special Civil Engineer Examination*, which contains two complete 2½ hour practice tests.

After solving these problems, you will be better able to identify your strengths and weaknesses in all subject areas of the Engineering Surveying Test Plan, which will help you make an informed decision regarding further review and preparation for the California Special Civil Engineer Examination. When you concentrate on strengthening your weak areas, you will be far better prepared and will improve your performance on the official exam.

Engineers who have taken the civil surveying exam frequently comment that there is not enough time to complete the questions. It is therefore imperative that you improve your problem-solving speed by becoming proficient in the use of a calculator—particularly with respect to computations involving angles. Angles, bearings, and distances are fundamental to virtually all survey applications. Unfortunately, most calculators do not easily handle degrees, minutes, and seconds, and the default angular unit is decimal degrees. You should make a point of learning the functions that enable rapid computation of trigonometric functions as well as addition, subtraction, and multiplication of angles directly in degrees, minutes, and seconds.

Pay special attention to the following topics, which are fundamental to surveying and form a necessary core of knowledge: coordinate geometry (COGO) functions (such as inverse, side shot, bearing-bearing intersection, traverse, areas), differential leveling, horizontal and vertical curves, earthwork volumes, and datums.

About the Exam

The California Special Civil Engineer Examination is given twice a year, in April and October. It tests the entry-level competency of a candidate to practice civil engineering within the profession's acceptable standards for public safety. This exam is open book and is administered over a 2½ hour period. The exam contains 50 multiple-choice problems derived from content areas as outlined in the board-adopted Engineering Surveying Test Plan and distributed as follows.

Equipment and Field Activities	20%
Field Measurements	10%
Calculations	40%
Office Procedures	30%

For each question, you will be asked to select the best answer from four choices. You'll record your answers on a machine-scorable answer sheet that will be provided to you when the exam is administered. Your calculations should be performed in the official test book, and they will not be credited or scored. Also, answers marked in the official test book will not be scored, and additional time will not be permitted to transfer answers to the official machine-scorable answer sheet.

The point values of each exam question are printed in the official test booklet. Points are assigned depending on the significance, difficulty, and complexity of the question.

What to Take to the Exam

You should come to the exam prepared with the following materials. Be sure to set them out ahead of time and devise a convenient way to carry them into the examination room with you.

- admission notice
- photo ID
- reference materials (only as much as you can carry in one trip)
- non-QWERTY keyboard calculator (it must not have communicating or text-editing capability)
- spare calculator batteries
- scales
- triangles
- compass
- protractor

Test-Taking Strategy

The California Special Civil Engineer Examination in surveying is a difficult test and requires thorough preparation in all areas, including multiple-choice test-taking techniques. Easy and difficult questions, with variable point values, are distributed throughout the exam. Besides contending with the nature and difficulty of the exam itself, many examinees spend too much time on difficult problems and leave insufficient time to answer the easy ones. You should avoid this. The following system can be very beneficial to you.

step 1: Work on the easy questions immediately and record your answers on your official machine-scorable answer sheet.

step 2: Work on questions that require minimal calculations and record your answers on your official machine-scorable answer sheet.

step 3: When you get to a question that looks "impossible" to answer, mark a "?" next to it in your

official test booklet, and go ahead and guess. Mark your "guess" answer on your answer sheet and continue.

step 4: When you face a question that seems difficult but solvable, mark an "X" next to it in your official test booklet; you may need considerable time to search for relevant information in your books, references, or notes. Continue to the next question.

step 5: When you come to a question that is solvable but you know requires lengthy calculations or is time consuming, mark a "+" sign next to the question in your official test booklet. For this question, you know exactly where to look for relevant information in your books, references, or notes.

Based on the number of exam questions and allotted time, on the average you should not spend more than 2.5 minutes per question. Thus, a lengthy or "time-consuming" question is one that will take you more than 2 minutes and 30 seconds to answer. Quickly and confidently, decide whether a question should receive a "+" or an "X;" the intent of this test-taking strategy is to save you precious time.

After you have gone over the entire exam, your official test booklet will clearly show which questions you have already answered and those that still require your attention.

You should now go over the exam a second time, with the following approach in mind.

step 1: The best and most successful approach is to go back and tackle: (a) the "+" questions, (b) the "X" questions, and (c) the "?" questions. As you proceed, eliminate your pluses, Xs, and question marks.

step 2: Recheck your work for careless mistakes.

step 3: Set aside the last few minutes of your exam period to fill in a guess for any unanswered lines on your official machine-scorable answer sheet. There is no penalty for guessing. Only questions answered correctly will be counted toward your score.

Exam Scoring

For the California Special Civil Engineer Examination in surveying, the board-adopted test plan lists the content areas of the exam and their assigned scoring percentages. The percentages assigned to each content area are the approximate proportion of total test points; however, the test plan does not reveal the total test points in advance (it varies from exam to exam). This makes it difficult to anticipate the exact number of problems for each content area of the exam.

The official exam is graded against a "cut score"—a predetermined minimum passing score that varies from exam to exam. Historically, if you score above 60% of the total examination point value, you have a chance of passing.

On the official exam, after initial scoring, any problem that the board finds to be flawed may be deleted. In the event of deletion, the point value of the deleted problem becomes zero and the total number of points possible on that exam is adjusted accordingly.

Nomenclature

A	area	in^2, yd^2
Az	azimuth	deg
b	base	ft
BC	beginning of curve	ft
BM	benchmark	ft
BS	backsight	ft
C	correction	ft
C	cost	$
d	diameter	in
d	distance	ft
D	degree of curvature	deg
D	distance	ft
dep	departure	ft
d_x	departure difference	ft
d_y	latitude difference	ft
e	error	ft
e	misclosure	ft
E	shortest distance from PI to circular curve	ft
EC	end of curve	ft
e_C	vector misclosure	ft
e_H	leveling misclosure	ft
e_x	x misclosure	ft
e_y	y misclosure	ft
elev	elevation of a point	ft
f	camera focal length	in
FS	foresight	ft
g	grade	%
h	height	ft
H	flying height above terrain	ft
HD	horizontal distance	ft
HI	height of instrument	ft
I	deflection angle	deg
K	interval factor	–
l	arc length	ft
L	length	ft
L	length of curve	ft
L	measured distance	ft
lat	latitude	ft
P	tension or pull	lbf
r	rate of change of grade	%/ft, %/sta

r	radius	ft
f	radius of a curve	ft
P	pull on tape	lbf
RR	rod reading	ft
S	distance between sections	ft
S	section	ft
SD	slope distance	ft
sta	station	sta, ft
T	tangent distance	ft
T	temperature	°F
V	volume	yd^3
VA	vertical angle	deg
v	residual	ft
w	weight per unit length of tape	lbf/ft
x	offset	ft

Symbols

α	angle	deg
α	angular resolution	deg
α	coefficient of expansion of steel	ft/ft-°F
β	angle	deg
β	slope of line (elevation angle)	deg
Δ	central angle	deg
θ	angle	deg
σ	standard error of a measurement	–
χ	angle	deg

Subscripts

BC	beginning of curve
BM	benchmark
BVC	beginning of vertical curve
e	earth
E	east or easting
EC	end of curve
HP	highest point
m	micrometer
N	north
PI	point of intersection
PVC	parabolic vertical curve
PVI	point of vertical intersection
s	sag (or unsupported), section, or segment
t	triangle

Problems

Equipment and Field Activities

1 The preferred method of obtaining the coordinates of a point P that is approximately 500 ft from two known points A and B in an area of flat terrain is to

- (A) take a GPS measurement of the baseline AP using survey-grade GPS receivers
- (B) take a total station side shot (bearing and distance) from B to P
- (C) compute a bearing/bearing intersection from observed azimuths AP and BP
- (D) take two side shots, AP and BP, using a steel tape and a theodolite

2 In the compass rule of traverse adjustment, the correction computed for each course as a result of balancing the traverse is a function of the

- (A) distance of the course and total transverse distance of the traverse
- (B) coordinate differences of the course
- (C) square root of the course distance
- (D) sum of the squares of all the course distances

3 In the adjustment or balancing of a closed differential leveling line, the amount of correction for an elevation is proportional to the

- (A) elevation difference between the measured point and the starting point
- (B) distance from the starting point to the measured point
- (C) number of setups between the starting point and the measured point
- (D) number of points between the starting point and the measured point

4 The scale of a stereo overlap formed from two aerial photos is determined from surveyed ground control. When redundant control is provided in the form of a third planimetric control point, the scale is determined from

- (A) the longest side between two control points
- (B) a least-squares solution using all three control points
- (C) the mean computed from three distances between control points
- (D) a weighted mean computed from three distances between control points

5 Control surveys for photogrammetric mapping should position the control points according to

- (A) the number and positions of photos and the contour interval
- (B) the nature of the terrain and the location of the high points
- (C) visibility of adjacent traverse points or GPS satellites
- (D) site accessibility

6 A standard total station has the ability to measure angles and distances electronically. A robotic total station has the additional function of

- (A) recording GPS data from satellites
- (B) directing a robot that records data automatically
- (C) automatically tracking the person carrying the prism reflector
- (D) automatically computing its latitude and longitude

7 A tilting level has a micrometer with which to adjust the split bubble. The field of view has three crosshairs, which can be used for standard stadia measurements using a stadia interval factor of 100 for the computation of distance. The instrument is leveled, and the top, middle, and bottom crosshair rod readings are 6.910 ft, 6.160 ft, and 5.410 ft, respectively. The micrometer is moved through one division, and the mid-crosshair reading is 6.145 ft. The angular resolution of the tilting level as represented by the micrometer scale in seconds of arc is most nearly

    (A) 5.0″
    (B) 10″
    (C) 15″
    (D) 21″

8 The rod used for leveling of the highest precision is made of

    (A) well-seasoned wood
    (B) an invar strip
    (C) stainless steel
    (D) an aluminum alloy

9 The minimum number of satellites required to obtain a position on the earth's surface is

    (A) 3
    (B) 4
    (C) 5
    (D) 8

10 Civil Engineers registered after January 1, 1982 are NOT authorized to perform

    (A) contour surveys using photogrammetry
    (B) geodetic surveying
    (C) surveys of fixed works
    (D) tunnel surveys

11 On a four-course open traverse ABCDE, the observed azimuth AB is 83°56′10″. The measured angles at points B, C, and D are 156°14′00″, 220°45′40″, and 130°29′30″, respectively.

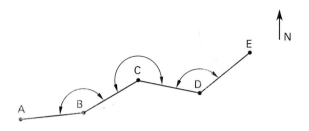

The bearing DE is

    (A) S 51°25′20″ W
    (B) N 51°25′20″ W
    (C) N 51°25′20″ E
    (D) N 231°25′20″ E

12 A closed traverse is preferred to a radial traverse because

    (A) points on a radial traverse are not checked
    (B) a closed traverse is more accurate
    (C) a closed traverse involves fewer observations
    (D) points are more easily visible on a closed traverse

13 An instrument designed to measure horizontal and vertical angles that are read from vernier scales is called a

    (A) theodolite
    (B) transit
    (C) total station
    (D) precise level

14 A CORS point is a

    (A) permanently recording GPS station
    (B) coordinated reference point surveyed by traditional triangulation
    (C) permanent benchmark on the national network
    (D) coordinated point that has been re-surveyed

15 Backsights and foresights should be equal in distance in order to

(A) minimize reading errors
(B) compensate for collimation errors
(C) compensate for mislevelment of the instrument
(D) eliminate errors due to a nonvertical rod

16 At a river crossing where it is impossible to equalize the backsight and foresight, which of the following leveling methods should be used?

(A) profile
(B) barometric
(C) trigonometric
(D) reciprocal

17 A control survey could be defined as one that surveys

(A) by using a level to control the elevations of cut/fill on an alignment
(B) radial measurements from a single base station, controlling the setting out of engineering structures
(C) a network of points covering a site, for the future control of additional survey measurements
(D) completed earthworks that act as a control on the final earthwork quantities

18 When setting up an automatic level, one should

(A) level approximately, using the circular bubble, and then precisely, using the single sensitive bubble
(B) set up the tripod and then let the automatic leveling take over
(C) level approximately, using the circular bubble
(D) level using the long sensitive bubble, and then turn the level approximately 90° and level again

19 The most accurate method of prolonging a straight line with a theodolite uses the method of

(A) double transit of the instrument
(B) turning the instrument through 180°
(C) single transit of the instrument
(D) turning the instrument through 180°, first clockwise, then counterclockwise

20 A total station has the ability to

(A) record angles electronically
(B) record distances electronically
(C) link to satellites
(D) both (A) and (B)

21 On a horizontal curve with a radius of 2400 ft and an intersection angle of $15°30'00''$, the distance from the point of intersection (PI) to the nearest point on the curve is

(A) 20.71 ft
(B) 21.92 ft
(C) 22.12 ft
(D) 22.21 ft

22 The following are the five interior observed angles of a traverse.

1	$139°56'10''$
2	$120°45'20''$
3	$141°08'50''$
4	$89°10'10''$
5	$48°58'40''$

Balanced angle 2 is

(A) $120°44'30''$
(B) $120°45'10''$
(C) $120°45'30''$
(D) $120°46'10''$

23 The accuracy of distance measurement for a total station is 0.01 ft ± 5 ppm. If the length of the measured line is 3867.4 ft, the accuracy of the measurement is most nearly

    (A) 0.010 ft
    (B) 0.019 ft
    (C) 0.030 ft
    (D) 0.20 ft

24 A map that is used to divide a parcel into five or more smaller parcels is called

    (A) a parcel map
    (B) a tentative map
    (C) a final map
    (D) an ALTA map

25 Civil engineers registered after January 1, 1982 are permitted to

    (A) determine the configuration or contour of the earth's surface or the positions of fixed objects thereon by means of measuring lines and angles
    (B) determine the position of any reference point or monument that defines a property line
    (C) perform geodetic or cadastral surveying
    (D) make any survey for the subdivision of land

26 The radius of the circular curve shown is 1800 ft, and the deflection angle to point P on the curve, $I/2$, is $3°45'00''$.

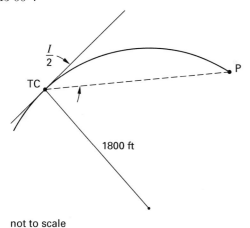

not to scale

The length of the curve from TC to P is most nearly

    (A) 210.77 ft
    (B) 229.34 ft
    (C) 235.62 ft
    (D) 261.80 ft

27 In construction staking, a 4 ft stake showing information relating to another stake is called

    (A) a witness stake
    (B) a hub
    (C) an offset stake
    (D) a lath

28 The initial bearing of line PQ on the three-course open traverse shown is N 85°30′ E. Angles α and β are $15°27'00''$ and $13°58'00''$, respectively.

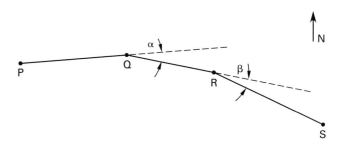

The azimuth of line RS is

    (A) 56°05′
    (B) 86°59′
    (C) 114°55′
    (D) 294°55′

29 Radial traverse PQ, PR, PS, PT has observed angles α, β, and θ, as shown.

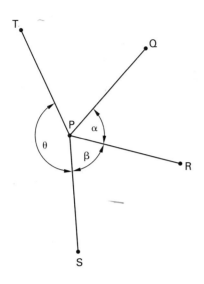

The bearing of PQ is N 41°57'20" E, and the observed angles are as follows.

$$\alpha = 62°10'10''$$
$$\beta = 71°44'30''$$
$$\theta = 158°32'40''$$

The bearing of PT is

(A) N 25°35'20" E
(B) N 25°35'20" W
(C) N 35°35'20" W
(D) N 64°24'40" W

30 On a construction staking project, the instrument that can be operated by a one-man crew is called

(A) an RTK GPS
(B) a total station
(C) a robotic total station
(D) both (A) and (C)

Field Measurements

31 For the points shown,

(x, y) coordinates of point L = (2000 ft, 5000 ft)
(x, y) coordinates of point P = (1400 ft, 5500 ft)
bearing LM = S 85°25'39" W
distance LM = 276.98 ft

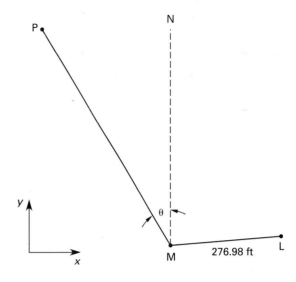

The bearing PM is

(A) N 31°48'56" W
(B) S 31°48'56" E
(C) N 58°11'04" W
(D) S 58°11'04" E

32 A line PQ is measured and recorded as a slope distance of 257.56 ft with a slope of 4°00'00". If the actual slope of the line is 3°00'00", the error introduced as a result of the incorrect slope is

(A) 0.16 ft
(B) 0.28 ft
(C) 0.96 ft
(D) 1.28 ft

33 A 100 ft steel tape is suspended in the air between a theodolite and a vertical rod, which is placed exactly at the 100 ft mark. The slope of the measured line is 1°40'. The tension on the tape is 18 lbf, and the weight of the tape per unit length is 0.02 lbf. The horizontal distance between the theodolite and the rod is

(A) 99.00 ft
(B) 99.91 ft
(C) 99.96 ft
(D) 100.01 ft

34 A surveyor measures a distance as 39.56 ft using break chaining. The tape is not accurately leveled and, at the plum bob end, is 2 ft lower than it should be, as shown.

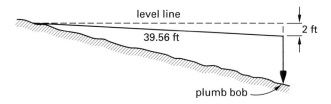

level line

39.56 ft

2 ft

plumb bob

The correct horizontal distance is

(A) 39.46 ft
(B) 39.51 ft
(C) 39.61 ft
(D) 39.66 ft

35 A horizontal angle is measured five times, and the results are as follows.

$$156°45'34'' \quad 156°45'38''$$
$$156°45'29'' \quad 156°46'30''$$
$$156°45'32''$$

The standard error of the measurement is most nearly

(A) 3″
(B) 4″
(C) 8″
(D) 14″

36 The mean direction between two azimuths, AB 289°30′50″ and AC 15°16′20″, expressed as a bearing is

(A) S 27°36′20″ E
(B) S 42°52′40″ E
(C) N 27°36′25″ W
(D) N 27°36′20″ E

37 In the "two-peg test" in leveling, backsight and foresight readings are taken at a setup midway between two points P and Q. Then, both points are sighted from an instrument set up outside of line PQ, and the results are used to determine the

(A) errors on the rods
(B) reading error of the surveyor
(C) mislevelment of the line of sight of the level
(D) accuracy of the circular bubble

38 On a closed differential leveling loop, the sum of the backsights minus the sum of the foresights equals

(A) zero
(B) the difference between the highest and lowest points on the loop
(C) an arbitrary number depending on the steepness of the terrain
(D) twice the value of the difference between the highest and lowest points of the loop

39 In a differential leveling line that comprises all backsights and foresights and that starts and ends on points with known elevations, the misclosure is adjusted into the intermediate points by

(A) an adjustment that is a function of the elevation of the point
(B) zero adjustment (the misclosure in leveling is usually extremely small and can be ignored)
(C) an adjustment proportional to the square of the distance
(D) a linear adjustment proportional to the distance

40 Consider a horizontal distance, PR, of 50.00 ft. The profile leveling field data for PQ is shown in the table.

point	BS	FS
P	4.67	–
Q	–	5.78
R	–	7.22

The grade of course PR is

(A) −5.1%
(B) −2.9%
(C) 2.9%
(D) 5.1%

41 A level line 350 ft in length is measured with a steel tape. A surveyor records the thermometer reading as 37° instead of 73°. The error in the length due to the reading error is

(A) 0.06 ft
(B) 0.08 ft
(C) 0.10 ft
(D) 0.11 ft

42 A 100 ft steel tape is standardized at P_0, a 15 lbf pull. The cross-sectional area of the tape is 0.003 in^2, and the elastic modulus of steel, E, is 3.0×10^7 lbf/in^2. If the actual tension on the tape, P, is 35 lbf and the distance reading, L, is 99.51 ft, the true distance is

(A) 99.49 ft
(B) 99.53 ft
(C) 99.55 ft
(D) 99.71 ft

43 A theodolite is set up at a point P, and a vertical angle is measured to the top of an inaccessible tower, T. The details of the measurement are

vertical (elevation) angle = 15°45′50″
height of instrument = 5.04 ft
(x, y) coordinates of P = (500.67 ft, 876.45 ft)
(x, y) coordinates of T = (710.06 ft, 1005.34 ft)
elevation of P = 1400.67 ft

The elevation of the tower is

(A) 1467.47 ft
(B) 1470.08 ft
(C) 1472.51 ft
(D) 1475.12 ft

Calculations

44 Side PQ of the triangle shown was measured with a steel tape as 174.98 ft. Angles P and Q are 58°14′10″ and 87°34′00″, respectively.

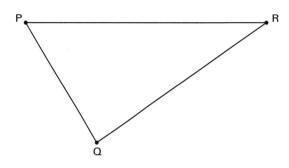

The length of side QR is

(A) 264.02 ft
(B) 264.70 ft
(C) 264.99 ft
(D) 265.46 ft

45 Sides a, b, and c of the triangle shown were measured as 330.56 ft, 210.90 ft, and 380.02 ft, respectively.

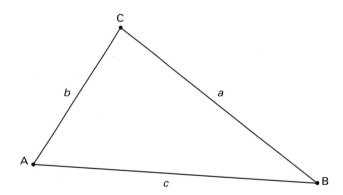

Angle B is

(A) 33°26′05″
(B) 33°27′33″
(C) 33°37′23″
(D) 34°02′30″

46 A theodolite at point A records an elevation angle of 22°30'00" to the top of a pole, P, as shown. A second theodolite at point B records an elevation angle of 42°20'00" to the same point, P. The horizontal distance AB is 321.00 ft. The theodolite at A is 19.65 ft higher than the theodolite at B and has an elevation of 500 ft above sea level. A, B, and P lie on a straight line (are in the same vertical plane).

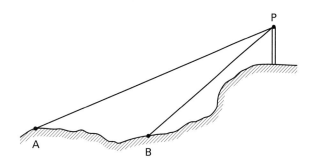

The elevation of the top of the pole above sea level is

(A) 699.84 ft
(B) 706.88 ft
(C) 727.09 ft
(D) 760.21 ft

47 For the triangle shown, course LM bears N 13°20' 40" W, and course LN bears N 39°59'10" E. Side n is 1298 ft and side m is 1487 ft.

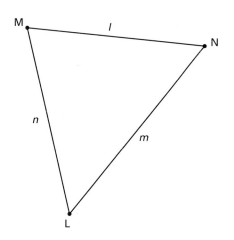

The area of the triangle is most nearly

(A) 576,300 ft^2
(B) 578,200 ft^2
(C) 774,100 ft^2
(D) 1,152,700 ft^2

48 A horizontal curve has a degree of curvature of 2°30'00". The stationing at the point of intersection is sta 23+44.78. The deflection angle of the curve at the PI is 12°00'00". The stationing at the beginning of curve (BC) is

(A) sta 18+57.64
(B) sta 21+03.90
(C) sta 25+85.66
(D) sta 28+25.66

49 A horizontal curve has a degree of curvature of 2°30'00". The deflection angle of the curve at the point of intersection, I, is 12°00'00". The stationing at the beginning of curve is sta 21+03.90. If points are staked every 100 ft on the curve, the deflection angle at the end of curve from the tangent to the second point on the curve will be

(A) 3°30'59"
(B) 3°32'56"
(C) 4°47'56"
(D) 7°05'51"

50 A curve has the characteristics shown.

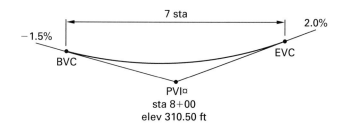

The elevation of sta 7+10 is

(A) 311.77 ft
(B) 312.09 ft
(C) 313.54 ft
(D) 321.34 ft

51 A curve has the characteristics shown.

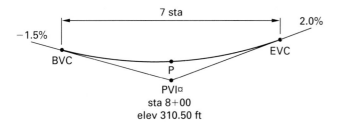

The elevation of the lowest point, P, on the curve is

(A) 313.50 ft
(B) 322.50 ft
(C) 323.67 ft
(D) 325.01 ft

52 An 18 in length of rebar is driven into the ground until it touches an underground pipe. A backsight to a benchmark with an elevation of 390.66 ft is taken with a tilting level. The backsight has a rod reading of 7.56 ft. A foresight is taken to a rod resting on top of the rebar. The rod reading is 1.01 ft. The elevation of the top of the pipe is

(A) 382.61 ft
(B) 385.95 ft
(C) 395.71 ft
(D) 398.71 ft

53 The measurements for a site are given in the following table.

point	BS (ft)	FS (ft)	elevation (ft)
A	3.34	–	1000.00
B	5.00	2.89	–
C	4.78	3.04	–
D	–	1.11	1006.18

The height of instrument, HI, at the setup between points B and C is

(A) 994.55 ft
(B) 1000.45 ft
(C) 1002.41 ft
(D) 1005.45 ft

54 The measurements for a site are given in the following table.

point	BS (ft)	FS (ft)	elevation (ft)
A	3.34	–	1000.00
B	5.00	2.89	–
C	4.78	3.04	–
D	–	1.11	1006.18

The misclosure at point D is

(A) −0.10 ft
(B) −0.08 ft
(C) +0.08 ft
(D) +0.10 ft

55 A surveyor measures a line 2.86 mi in length. The combined earth curvature and refraction correction for the line is

(A) 2.11 ft
(B) 4.70 ft
(C) 5.10 ft
(D) 6.20 ft

56 In the cross section shown,

$$\text{elevation of point T} = 760.00 \text{ ft}$$
$$\text{HI above point T} = 4 \text{ ft } 7 \text{ in}$$
$$\text{elevation angle} = -7°15'50''$$
$$\text{rod reading} = 3.99 \text{ ft}$$
$$\text{horizontal distance TP} = 320.76 \text{ ft}$$

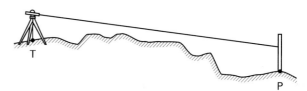

not to scale

The elevation of point P is

(A) 719.71 ft
(B) 727.69 ft
(C) 801.47 ft
(D) 809.45 ft

57 A backsight to a benchmark at a point A reads 2.55 ft. The elevation of point A is 2800.20 ft above mean sea level. A surveyor is required to place a mark on a lath, which would have an elevation of 2798.11 ft. The rodman holds the rod next to the lath. To set the base of the rod to this elevation, the rod reading would be

    (A) 0.56 ft
    (B) 0.76 ft
    (C) 4.46 ft
    (D) 4.64 ft

58 An engineer has to measure the distance PR despite the presence of an obstruction. The (x, y) coordinates $(500 \text{ ft}, 1000 \text{ ft})$ are assumed for point P, and the engineer traverses around the obstruction via point Q.

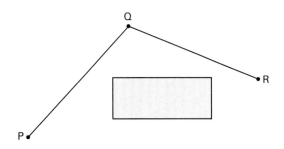

The traverse measurements are given in the following table.

course	distance, D (ft)	azimuth, Az
PQ	151.84	$43°16'00''$
QR	145.22	$111°29'00''$

Distance PR is

    (A) 230.77 ft
    (B) 245.99 ft
    (C) 250.91 ft
    (D) 258.15 ft

59 A five-course traverse has the following measured distances: 1367.2 ft, 2756.0 ft, 1163.8 ft, 998.3 ft, and 3672.5 ft. The misclosures in x and y are 0.54 ft and 0.23 ft, respectively. The precision expressed as a fraction is most nearly

    (A) $\dfrac{1}{12,900}$

    (B) $\dfrac{1}{17,000}$

    (C) $\dfrac{1}{18,400}$

    (D) $\dfrac{1}{28,900}$

60 A survey specification requires a maximum misclosure of 1/80,000 on all traversing. The project area measures 1.6 mi by 2.4 mi. A perimeter traverse is run around the project. The allowable misclosure on this traverse is

    (A) 0.50 ft
    (B) 0.53 ft
    (C) 0.58 ft
    (D) 0.64 ft

61 For the points shown,

$$\text{bearing BP} = \text{N } 30°00'00'' \text{ W}$$
$$\text{angle } \alpha = 92°10'00''$$
$$(x, y) \text{ coordinates of point A} = (500 \text{ ft}, 1100 \text{ ft})$$
$$(x, y) \text{ coordinates of point B} = (1000 \text{ ft}, 1000 \text{ ft})$$

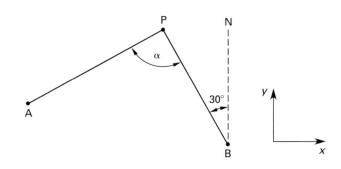

The (x, y) coordinates of point P are

(A) $(-873.49 \text{ ft}, -1278.96 \text{ ft})$
(B) $(322.05 \text{ ft}, 1006.05 \text{ ft})$
(C) $(838.72 \text{ ft}, 1279.34 \text{ ft})$
(D) $(838.94 \text{ ft}, 1278.96 \text{ ft})$

62 For the circular arc shown, the (x, y) coordinates of the center, C, are $(5000 \text{ ft}, 5000 \text{ ft})$.

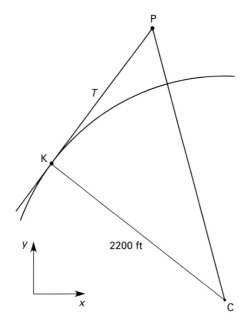

If the (x, y) coordinates of point P are $(4400 \text{ ft}, 7700 \text{ ft})$, the tangent distance, T, is

(A) 1587.43 ft
(B) 1676.30 ft
(C) 2887.30 ft
(D) 3177.46 ft

63 The geoid is

(A) the geodetic reference surface (a rotated ellipse) that is the basis of the state-plane coordinate system
(B) the deviation of mean sea level from the reference spheroid
(C) the mean sea-level surface covering the entire globe
(D) the actual surface of the earth both above and below sea level

64 A triangular ditch 120 ft in length and 30 ft wide is cut over flat terrain. The depth of the ditch is measured as 15.9 ft at one end and 18.94 ft at the other end. There is a uniform slope along the bottom of the ditch. The volume of earth removed from the ditch is most nearly

(A) 927 yd³
(B) 1160 yd³
(C) 2230 yd³
(D) 3340 yd³

65 Dirt from a hole 25 ft in diameter and 10.5 ft deep is excavated and moved 800 ft. If the cost of excavation is \$2.50/yd³ and the cost of overhaul is \$0.60/yd³ per station with a freehaul of four stations, the total cost of the earthworks is most nearly

(A) \$300
(B) \$680
(C) \$930
(D) \$3740

66 The California State Plane System of 1983 (NAD83) is based on which of the following ellipsoids?

(A) Clarke spheroid of 1866
(B) Fisher spheroid of 1960
(C) IUGG spheroid of 1967
(D) Geodetic Reference System of 1980

67 The State Plane System divides California into several regions or "zones." How many zones are associated with the NAD 27 system?

(A) 2
(B) 5
(C) 6
(D) 7

68 A 2400 ft radius curve has a point of curvature with a station of sta 18+18.99. Points are set on the curve at half stations. The tangent offset for the third point on the curve is

(A) 1.37 ft
(B) 2.73 ft
(C) 3.57 ft
(D) 7.15 ft

69 A two-course traverse is measured with sides AB and BC. The starting point, A, has (y, x) coordinates $(1000.00 \text{ ft}, 500.00 \text{ ft})$, and the measured distances and azimuths are

side	distance, D (ft)	azimuth, Az
AB	267.56	319°54′20″
BC	189.55	18°44′30″

Point C has known (y, x) coordinates $(1384.38 \text{ ft}, 388.28 \text{ ft})$. The misclosures of the traverse are

(A) −0.20 ft, +0.30 ft
(B) +0.30 ft, −0.20 ft
(C) +0.52 ft, −0.10 ft
(D) +0.61 ft, +0.30 ft

70 A benchmark, BMc, has an elevation of 763.60 ft, and the horizontal distance from BMb to BMc is 200.00 ft. Backsights and foresights are given in the table.

point	BS (ft)	FS (ft)	elev (ft)
BMa	4.67	–	783.67
	3.00	5.59	–
	4.26	6.51	–
	3.22	5.74	–
BMb	–	7.38	–
BMc	–	–	763.60

The ground between BMb and BMc has a uniform grade. The grade of the line from BMb to BMc is

(A) −5%
(B) −2%
(C) 2%
(D) 5%

71 The trigonometric leveling observations for a site are as shown.

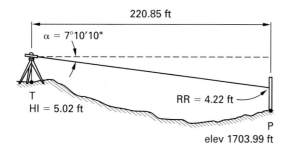

not to scale

The ground elevation of T is

(A) 1677.01 ft
(B) 1730.75 ft
(C) 1730.97 ft
(D) 1731.77 ft

72 The U.S. survey foot was introduced with the implementation of the

(A) North American Datum of 1927
(B) North American Datum of 1983
(C) National Geodetic Vertical Datum of 1929
(D) North American Vertical Datum of 1988

73 On the 2000 ft radius curve shown, the tangent distance is 278.346 ft, and the tangent offset is 19.464 ft.

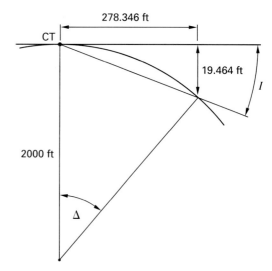

not to scale

Determine the central angle, Δ.

(A) 2°
(B) 4°
(C) 6°
(D) 8°

74   Sections A and B are part of a highway design with section spacing of 100 ft and a cross section as shown.

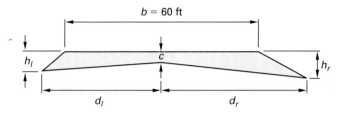

section	h_l (ft)	h_r (ft)	d_l (ft)	d_r (ft)	c (ft)
A	5.81	6.44	38.91	42.98	3.01
B	5.76	7.94	37.00	44.95	3.21

The volume of the 100 ft section AB is most nearly

(A) 180 yd^3
(B) 360 yd^3
(C) 1190 yd^3
(D) 2390 yd^3

75   Trigonometric leveling does not have the inherent high precision of standard backsight/foresight leveling because of

(A) the earth curvature correction
(B) the inaccuracy of measuring the elevation angle
(C) refraction of the observed line through the atmosphere
(D) the inaccuracy of the distance measurement

76   Backsight, foresight, and elevation data for two points, P and Q, are given in the table.

point	BS (ft)	FS (ft)	elevation (ft)
P	4.68	–	1000.00
	3.99	2.12	–
	2.07	1.63	–
	5.55	1.64	–
Q	–	4.71	–

The elevation of point Q is

(A) 993.81 ft
(B) 1003.98 ft
(C) 1006.19 ft
(D) 1015.61 ft

77   The following two-course traverse was measured in the field.

course	distance, D (ft)	azimuth, Az
AB	267.45	127°03′20″
BC	180.99	190°30′30″

If the (x, y) coordinates of A are $(900.00 \text{ ft}, 850.00 \text{ ft})$, the y coordinate of point C is

(A) 510.88 ft
(B) 560.88 ft
(C) 1189.12 ft
(D) 1239.12 ft

78   A reverse curve has the characteristics shown.

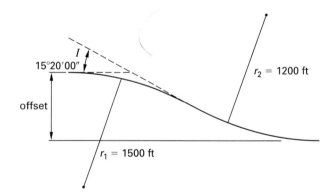

not to scale

The offset is

(A) 10.68 ft
(B) 79.33 ft
(C) 96.11 ft
(D) 713.97 ft

79 A triangle has sides measured as shown.

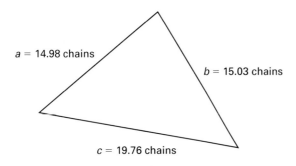

The area of the triangle is most nearly

(A) 9.86 ac
(B) 11.16 ac
(C) 15.78 ac
(D) 134.36 ac

80 Triangle ABC has sides measured as shown.

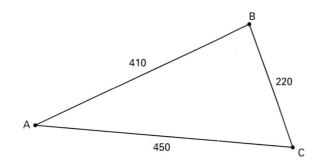

Angle A is

(A) 29°06′05″
(B) 29°10′15″
(C) 29°17′08″
(D) 29°20′00″

81 A circular curve is segmented as shown.

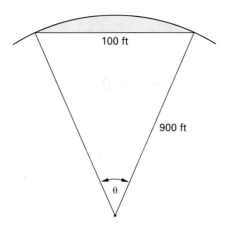

not to scale

The area bounded by the circular curve and the 100 ft line is most nearly

(A) 87 ft^2
(B) 92 ft^2
(C) 94 ft^2
(D) 98 ft^2

82 The traverse side AB shown has a latitude of −100 ft and a departure of −200 ft.

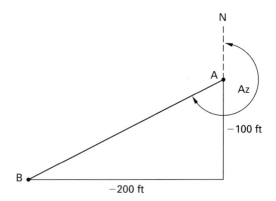

The azimuth of line AB is

(A) 63°26′06″
(B) 63°43′49″
(C) 243°26′06″
(D) 243°43′49″

83 Given that the (x, y) coordinates of two points, M and N, are $(100, -100)$ and $(500, -800)$, respectively, bearing MN is

(A) S 29°44′42″ E
(B) S 60°15′18″ E
(C) N 29°44′42″ W
(D) N 60°15′18″ W

84 For the points shown,

bearing CB = N 82°15′00″ E
bearing CA = S 40°00′00″ E

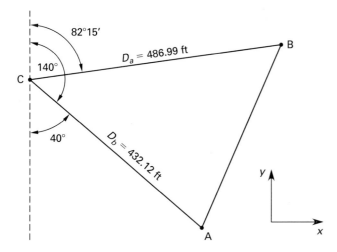

The area of triangle ABC is most nearly

(A) 56,100 ft²
(B) 89,000 ft²
(C) 112,000 ft²
(D) 178,000 ft²

85 The radii and central angles of a compound curve are 2000 ft, 15°05′00″ and 2450 ft, 11°45′00″. The total length of the compound curve is

(A) 1014.95 ft
(B) 1028.94 ft
(C) 1047.40 ft
(D) 1055.12 ft

86 The computed departure between two points, A and B, is 154.86 ft, and distance AB is 201.66 ft. The total traverse distance is 589.01 ft, and the easting traverse misclosure is 0.24 ft. The balanced departure of line AB using the compass rule is

(A) 154.78 ft
(B) 154.92 ft
(C) 155.01 ft
(D) 155.04 ft

87 A distance of 2856.11 ft is measured with a total station at an elevation of 5400 ft above sea level. The radius of the earth is 20,906,000 ft. The distance, corrected for the sea-level scale factor, is

(A) 2854.80 ft
(B) 2855.37 ft
(C) 2856.85 ft
(D) 2857.01 ft

88 For the section shown, sections S_1 and S_2 are spaced 100 ft apart.

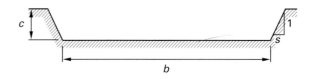

The dimensions of S_1 and S_2 are as given in the table.

section	b	c	s
S_1	40.0 ft	5.67 ft	0.5
S_2	50.0 ft	7.14 ft	0.5

The area of the midsection in a prismoidal computation of the volume of dirt removed is

(A) 308.99 ft²
(B) 312.68 ft²
(C) 315.87 ft²
(D) 420.99 ft²

89 The station of the point of intersection (PI) on a horizontal curve is sta 18+37.9. The length of the curve is 650.4 ft, and the tangent distance is 2.1009 sta. A point is placed at every full station on the curve. The distance from the tangent point to the first point on the curve is

    (A) 27.81 ft
    (B) 72.19 ft
    (C) 127.81 ft
    (D) 172.91 ft

90 A profile is measured as shown.

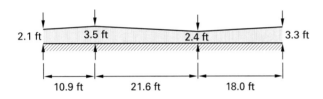

The area of the profile is

    (A) 130 ft^2
    (B) 150 ft^2
    (C) 290 ft^2
    (D) 300 ft^2

91 For a parabolic vertical curve,

$$\text{station of PVI} = \text{sta } 16+40.00$$
$$\text{elevation of PVI} = 744.41 \text{ ft}$$
$$\text{grade } 1 = +2\%$$
$$\text{grade } 2 = -3\%$$
$$\text{length of curve} = 7 \text{ sta}$$

The station of the highest point on the curve is

    (A) sta 10+10.00
    (B) sta 13+18.00
    (C) sta 15+70.00
    (D) sta 17+70.00

92 An existing highway is detoured due to an obstruction, as shown. A radius of 1000 ft must be held for each curve in a reverse curve.

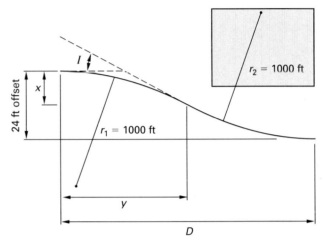

not to scale

The distance required to develop a 24 ft offset is

    (A) 77.23 ft
    (B) 154.45 ft
    (C) 308.90 ft
    (D) 439.09 ft

93 On a parabolic vertical curve,

$$\text{grade } 1 = -2\%$$
$$\text{grade } 2 = 1.5\%$$
$$\text{station of PVI} = \text{sta } 28+15.00$$
$$\text{elevation of PVI} = 644.73 \text{ ft}$$
$$\text{length of curve} = 8 \text{ sta}$$

The elevation of the first full station on the curve is

    (A) 643.12 ft
    (B) 646.68 ft
    (C) 651.19 ft
    (D) 654.68 ft

94 On course AB of a traverse, the departure is 249.87 ft and the latitude is −180.94 ft. Bearing AB is

(A) E 35°54′35″ S
(B) S 35°54′35″ E
(C) N 35°54′35″ E
(D) S 54°05′25″ E

95 The table shown refers to a five-course traverse. The known coordinates of point E are (185.37 ft, 358.00 ft).

course	distance (ft)	point	N (ft)	E (ft)
AB	156.45	A	100.00	100.00
BC	145.89	B	254.07	127.17
CD	122.22	C	357.23	230.33
DE	167.89	D	352.97	352.47
–	–	E	185.18	358.33

The balanced coordinate of point B is

(A) (254.12 ft, 127.08 ft)
(B) (254.02 ft, 127.26 ft)
(C) (254.00 ft, 127.28 ft)
(D) (254.15 ft, 127.05 ft)

96 The degree of curvature based on the arc definition of the circular curve shown is 2°. The distance from point P to the circle center is 4160.10 ft.

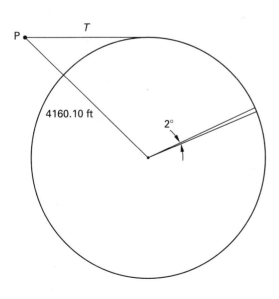

The tangent distance is

(A) 2980.47 ft
(B) 3010.99 ft
(C) 3016.52 ft
(D) 3033.71 ft

97 For the curve shown,

degree of curvature (arc definition)
of horizontal curve TC–P = 2°30′
length of curve TC–P = 145.00 ft
bearing TC–Q = S 66°14′40″ W

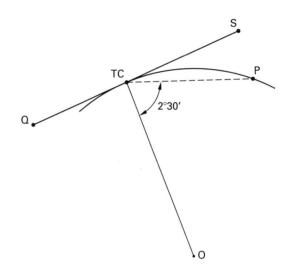

not to scale

Bearing TC–P is

(A) N 68°03′25″ E
(B) N 69°52′10″ E
(C) S 68°03′25″ W
(D) S 69°03′25″ W

98 A three-course traverse ABCD has the latitudes and departures given in the table.

course	latitude (ft)	departure (ft)
AB	+100	+100
BC	−30	+90
CD	−90	+20

The azimuth AD is

(A) 84°15′57″
(B) 84°33′35″
(C) 95°26′25″
(D) 95°44′03″

99 On a survey on NAD 83, the grid scale factor is 0.999967, the sea-level scale factor is 0.999351, and measured distance AB is 8456.9 ft. Distance AB on the state-plane grid is equal to

(A) 8451.1 ft
(B) 8451.7 ft
(C) 8462.1 ft
(D) 8462.7 ft

100 For the points shown,

(y, x) coordinates of L = (540.55, 879.01)
(y, x) coordinates of M = (638.78, 925.43)
bearing MP = due west
bearing LP = N 68°35′00″ W

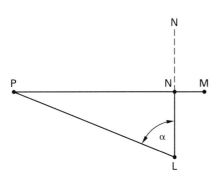

The (y, x) coordinates of point P are

(A) (638.78, 628.57)
(B) (628.57, 638.78)
(C) (638.78, 840.48)
(D) (638.78, 1129.45)

101 On a mass diagram, a rising curve indicates

(A) excavation
(B) embankment
(C) borrow
(D) waste

102 Surveys based on NAD 83 on the California state-plane coordinate system are placed on one of six zones. The state is divided into six zones because

(A) each geographical center maintains its own local zone
(B) the state is too large to accommodate a single coordinate system
(C) the six largest counties maintain the boundary records on their state plane zones
(D) the zones follow the California GIS, which is maintained in six separate data bases

Office Procedures

103 In the U.S. Public Lands System, the length of the side of a quarter section is

(A) 20 chains
(B) 40 chains
(C) 60 chains
(D) 80 chains

104 In the legal definition of property, the courts have usually ruled that the most important factor in the interpretation of property is the

(A) intent of the parties to a conveyance
(B) monuments on the ground
(C) field book measurements
(D) intent of the surveyor

105 The national leveling network that was adjusted to fit 26 mean sea level stations is

(A) NAVD 88
(B) NGVD 29
(C) NAD 83
(D) NAD 27

106 A benchmark is generally defined as a

(A) permanently marked point with a known elevation

(B) permanently marked point with a known latitude and longitude

(C) continuously recording GPS point

(D) point that defines mean sea level in a tidal area of water

107 Three contour lines are depicted, with points A and B positioned as shown.

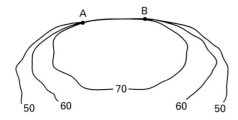

The convergence between points A and B represents

(A) an overhang

(B) a vertical cliff

(C) a ridge

(D) a physical impossibility

108 A photo has a scale of 1 in:300 ft. A water storage tank measures 8.5 mm on the photo. The size of the tank at ground scale is most nearly

(A) 8.4 ft

(B) 100 ft

(C) 200 ft

(D) 900 ft

109 Flight lines must be drawn on a 1 in:2000 ft USGS quad sheet for a photogrammetric mapping project using a photo scale of 1 in:400 ft. The side lap is 30%. The distance between the flight lines on the quad map is

(A) 0.38 in

(B) 0.75 in

(C) 1.26 in

(D) 1.51 in

110 In the Public Lands System, the length of the side of a quadrangle is

(A) 1.0 mi

(B) 6.0 mi

(C) 24 mi

(D) 48 mi

111 A forward overlap of 80% is required for a orthophoto mapping project. The orthophoto map scale is 1 in:100 ft, and the enlargement between the photo and the map is 8x. The distance between photo centers at ground scale is

(A) 640 ft

(B) 1440 ft

(C) 2880 ft

(D) 5760 ft

112 A legal description involving a complete perimeter description where each course (side) is described in sequence in a uniform direction of travel is known as a

(A) metes and bounds description

(B) bounds description

(C) subdivision description

(D) proportional conveyance

113 Full control on an aerial stereo overlap requires

(A) two points, with a known plan position and elevation

(B) three elevation points (one in each corner) and two plan points

(C) four points (one in each corner) with a known plan position and elevation

(D) six points (one in each corner and one at each photo center) with a known plan position and elevation

114 A photo image in which every pixel is in its correct map position is known as a

(A) photomap

(B) digital image

(C) digital orthophoto

(D) raster map

115 The national vertical datum that is not a sea-level datum and was held fixed to a tidal benchmark in Quebec is known as

(A) NAD 27
(B) CAN 91
(C) NGVD 29
(D) NAVD 88

116 A 6 in focal length camera is flown at 11,000 ft above sea level for an aerial survey project. The average terrain elevation is 5000 ft. The photo scale is most nearly

(A) 1 in:500 ft
(B) 1 in:1000 ft
(C) 1 in:1800 ft
(D) 1 in:11,000 ft

117 On a contour map of an area compiled from aerial photos using a stereoplotter, the contours are usually generated by

(A) directly drawing the contours by stereo measurement
(B) computer interpolation from a digital elevation model
(C) computer interpolation from profiles
(D) automatically generating contours by image-matching

118 On a parabolic vertical curve,

$$\text{grade } 1 = -2\%$$
$$\text{grade } 2 = +3\%$$
$$\text{length of curve} = 8 \text{ sta}$$
$$\text{station of PVI} = \text{sta } 17{+}56.00$$
$$\text{elevation of PVC} = 450.00 \text{ ft}$$

The station of the lowest point on the curve is

(A) sta 10+36.00
(B) sta 13+68.50
(C) sta 16+76.00
(D) sta 17+16.00

119 A surveyor locates the NW corner of section 15 and the SE corner of section 22 on the same township. The bearing from the section 22 corner to the section 15 corner is

(A) N 26°34′ W
(B) N 26°34′ E
(C) W 63°26′ N
(D) E 63°26′ N

120 The process of determining the ground coordinates of artificial points marked on the emulsion of an aerial photo is known as

(A) ortho rectification
(B) aerial triangulation
(C) terrestrial photogrammetry
(D) absolute orientation

Solutions

Equipment and Field Activities

1 *The answer is (D).*

The preferred method is option (D) for the simple reason that a check is provided by having two side shots. Even though use of a steel tape is considered inaccurate compared to other more modern methods, it is acceptable over a short distance such as 500 ft. The three other methods are unchecked. A fundamental rule of surveying is to check one's work as much as possible.

2 *The answer is (A).*

The compass rule computes a correction separately in x and y. The departure correction, for example, is $d_x = (D_{\text{course}}/D_{\text{total}})\, e_x$. The adjustment is therefore a function of the distance of the course and the total traverse distance.

3 *The answer is (D).*

The correction applied to point i in a closed leveling loop is
$$\left(\frac{n-i}{n}\right) e_H$$

n is the total number of points in the circuit.

The correction or adjustment is therefore proportional to the sequential number, i, of the point in the leveling circuit.

4 *The answer is (B).*

Surveying and mapping problems invariably involve redundant data in order to provide a check on measurements. When this occurs, the correct procedure would be to use the method of least squares to compute the unknowns.

5 *The answer is (A).*

Control points for aerial mapping projects are placed in predetermined positions. The spacing between the control points along each flight line is a function of the contour interval. Most of the control points are placed on the lateral overlap between adjacent flight lines. The control point positions are therefore determined by the number of photos and the contour interval.

6 *The answer is (C).*

The main feature of a robotic total station is the ability to track a prism automatically. The instrument is equipped with servo motors that turn the telescope both horizontally and vertically. The instrument stops turning when the laser beam hits the prism.

7 *The answer is (D).*

The horizontal distance from the instrument to the rod is
$$\begin{aligned} D &= K(\text{RR}_1 - \text{RR}_2) \\ &= (100)(6.910 \text{ ft} - 5.410 \text{ ft}) \\ &= 150 \text{ ft} \end{aligned}$$

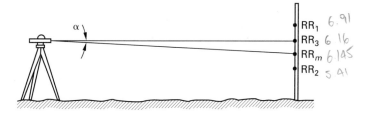

not to scale

The tangent of the angular resolution of the micrometer can now be found.

$$\tan\alpha = \frac{RR_3 - RR_m}{D}$$
$$= \frac{6.160\text{ ft} - 6.145\text{ ft}}{150\text{ ft}}$$
$$= 0.00010$$

Therefore,

$$\alpha = \tan^{-1} 0.00010$$
$$= 20.6'' \quad (21'')$$

8   *The answer is (B).*

Leveling of the highest precision is called geodetic or precise leveling. A geodetic leveling rod consists of an invar strip held in a casing of well-seasoned wood or an extruded aluminum alloy. The readings are marked on the invar strip.

9   *The answer is (B).*

The number of unknowns involved in a point position is three, namely latitude, longitude, and height above sea level (or more precisely, height above the ellipsoid). Three satellites would normally be required to obtain these unknowns. However, satellite atomic clocks and receiver clocks drift, and this time unknown requires a fourth satellite for its solution.

10   *The answer is (B).*

Options (A), (C), and (D) can be performed by any registered civil engineer. Geodetic surveys, however, are performed only by registered land surveyors or civil engineers who were registered before January 1, 1982.

11   *The answer is (C).*

The azimuths of BA and BC are

$$Az_{BA} = Az_{AB} + 180°00'00''$$
$$= 83°56'10'' + 180°00'00''$$
$$= 263°56'10''$$
$$Az_{BC} = Az_{BA} + \text{angle B}$$
$$= 263°56'10'' + 156°14'00''$$
$$= 420°10'10''$$

Since $Az_{BC} > 360°$, subtract 360° from the azimuth.

$$Az_{BC} = 420°10'10'' - 360° = 60°10'10''$$
$$Az_{CB} = Az_{BC} + 180°00'00''$$
$$= 60°10'10'' + 180°00'00''$$
$$= 240°10'10''$$
$$Az_{CD} = Az_{CB} + \text{angle C} - 360°$$
$$= 240°10'10'' + 220°45'40'' - 360°$$
$$= 100°55'50''$$
$$Az_{DC} = Az_{CD} + 180°00'00''$$
$$= 100°55'50'' + 180°00'00''$$
$$= 280°55'50''$$
$$Az_{DE} = Az_{DC} + \text{angle D} - 360°$$
$$= 280°55'50'' + 130°29'30'' - 360°$$
$$= 51°25'20''$$

Since an azimuth is measured clockwise from north, the bearing is N 51°25'20'' E.

12   *The answer is (A).*

A radial traverse would be more accurate than a closed traverse since each point is measured directly from the known central point instead of being part of a loop. A radial traverse involves one fewer observation than does a closed traverse. Intervisibility is not a function of the type of traverse but depends on the placement of points. Since a radial traverse point is coordinated by a single side shot (bearing and distance), it is essentially unchecked, as opposed to a point on a closed traverse, which is on a closed loop and is therefore checked.

13   *The answer is (B).*

A precise level does not measure angles but generates a level line of sight. A total station records angles electronically. A theodolite has glass circles from which angles are read optically. A transit predates the other three instruments mentioned and uses verniers to read horizontal and vertical angles.

14   *The answer is (A).*

California has a network of permanently recording GPS geodetic stations that provides a reference framework for surveyors and engineers. The data recorded on these stations is available on the internet and is frequently used on engineering projects.

15 *The answer is (B).*

Collimation errors occur when the instrument is properly leveled and yet the line of sight is not level. When this occurs and the backsight and foresight distances are equal, the resulting error on the rod reading is identical in magnitude for both the backsight and the foresight. Thus the errors cancel out during computation of the leveling. It is standard procedure to place the rod at equal backsights and foresights for virtually all types of leveling.

16 *The answer is (D).*

Errors due to collimation occur when the backsight and foresight distances are not equal. These errors do not occur in barometric or trigonometric leveling. Profile leveling is usually associated with leveling along a route of some length, not at a site such as a river crossing. In reciprocal leveling, collimation errors can be eliminated by computing the mean of the two elevation differences as observed from the two sides of the river.

17 *The answer is (C).*

A control survey is the setting out of a network of points that are used as a reference for future surveys. This network can cover an area as large as a state or as small as a few acres.

18 *The answer is (C).*

An automatic level does not have a sensitive bubble; it has only a circular bubble. Once an approximate leveling is achieved, the automatic device takes over and performs the final accurate leveling without operator intervention.

19 *The answer is (A).*

The most accurate method of prolonging a straight line with a theodolite is to sight back along the line and then transit the instrument and place a point on the extended line. Next, turn the instrument without transiting, sight the original back point, and transit and place a second point on the extended line at the same distance as the first point. This is known as a double transit. In general, the two points so placed will not coincide. The mean position selected from these two points will lie exactly on the extended straight line.

20 *The answer is (D).*

A total station is an electronic theodolite with the ability to measure distances using a laser beam. It does not have any GPS capability and, therefore, does not receive any satellite data. It does, however, record angles electronically and is capable of storing recorded data for later downloading to a PC.

21 *The answer is (C).*

The distance to the nearest point on the curve is illustrated by the distance E.

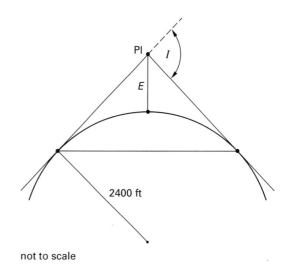

not to scale

Distance E can be found from

$$E = r \left(\sec \frac{I}{2} - 1 \right)$$
$$= (2400 \text{ ft}) \left(\sec \frac{15°30'00''}{2} - 1 \right)$$
$$= (2400 \text{ ft})(0.0092)$$
$$= 22.12 \text{ ft}$$

22 *The answer is (C).*

To find the misclosure, sum the observed angles.

$$\sum \text{observed angles} = 139°56'10'' + 120°45'20''$$
$$+ 141°08'50'' + 89°10'10''$$
$$+ 48°58'40''$$
$$= 539°59'10''$$

On an n-sided polygon,

$$\sum \text{angles} = (n-2)180°$$

Therefore,

$$e = \sum \text{angles} - \sum \text{observed angles}$$
$$= (n-2)180° - 539°59'10''$$
$$= 50''$$

Now find the correction.

$$C = \frac{e}{n} = \frac{50''}{n}$$
$$= \frac{50''}{5}$$
$$= 10'' \text{ per angle}$$

Balanced angle 2 is found to be

$$\text{balanced angle } 2 = \text{observed angle } 2 + C$$
$$= 120°45'20'' + 10''$$
$$= 120°45'30''$$

23 *The answer is (C).*

The accuracy of the total station is

$$\text{accuracy} = 0.01 \text{ ft} + 5\left(\frac{D}{10^6}\right)$$
$$= 0.01 \text{ ft} + (5)\left(\frac{3867.4 \text{ ft}}{10^6}\right)$$
$$= 0.03 \text{ ft}$$

24 *The answer is (C).*

When a parcel is subdivided into five or more parcels, it is known as a final map.

25 *The answer is (A).*

Only civil engineers registered before January 1, 1982 are permitted to determine the position of any reference point or monument that defines a property line, perform geodetic or cadastral surveying, or make any survey for the subdivision of land. However, any registered civil engineer can determine the configuration or contour of the earth's surface or the positions of fixed objects thereon by means of measuring lines and angles.

26 *The answer is (C).*

Determine the deflection angle, in radians.

$$I = 2\left(\frac{I}{2}\right) = (2)(3°45'00'')$$
$$= 7°30'00'' \quad (7.5°)$$

The length of curve is

$$L = rI = (1800 \text{ ft})(7.5°)\left(\frac{\pi}{180°}\right)$$
$$= 235.62 \text{ ft}$$

27 *The answer is (D).*

When a stake is set on a construction project, it is often accompanied by another stake called a lath (2 ft or 4 ft in length), which contains information about the stake, such as distance to the stake or elevation of the top of the stake.

28 *The answer is (C).*

Since line PQ is in the first quadrant,

$$\text{Az}_{PQ} = \text{bearing angle of PQ} = 85°30'00''$$

The azimuth of line QR is

$$\text{Az}_{QR} = \text{Az}_{PQ} + \alpha = 85°30' + 15°27'$$
$$= 100°57'$$

The azimuth of line RS can now be found.

$$\text{Az}_{RS} = \text{Az}_{QR} + \beta = 100°57' + 13°58'$$
$$= 114°55'$$

29 *The answer is (B).*

Since PQ lies in the first quadrant, its azimuth is the bearing angle, which is $41°57'20''$.

$$\begin{aligned}
Az_{PT} &= Az_{PQ} + \alpha + \beta + \theta \\
&= 41°57'20'' + 62°10'10'' + 71°44'30'' \\
&\quad + 158°32'40'' \\
&= 334°24'40''
\end{aligned}$$

The bearing angle of PT is

$$\begin{aligned}
\text{bearing angle of PT} &= 360° - Az_{PT} \\
&= 360° - 334°24'40'' \\
&= 25°35'20''
\end{aligned}$$

PT lies in the fourth quadrant; therefore, bearing PT is N $25°35'20''$ W.

30 *The answer is (D).*

An RTK GPS (real-time kinematic global positioning system) is linked to a base station via radio, and stakes can be set by the person operating the rover receiver. The base receiver does not require any human intervention. A total station is essentially a two-person field operation. A robotic total station can be operated by one person since the instrument has a self-seeking laser that can locate and bisect the rover prism. Thus, both RTK GPS and the robotic total station can be operated by one person.

Field Measurements

31 *The answer is (B).*

The azimuth of LM is

$$\begin{aligned}
Az_{LM} &= 85°25'39'' + 180° \\
&= 265°25'39''
\end{aligned}$$

The coordinates of point M are

$$\begin{aligned}
x_M &= x_L + D_{LM} \sin Az_{LM} \\
&= 2000.00 \text{ ft} + (276.98 \text{ ft}) \sin 265°25'39'' \\
&= 1723.90 \text{ ft} \\
y_M &= y_L + D_{LM} \cos Az_{LM} \\
&= 5000.00 \text{ ft} + (276.98 \text{ ft}) \cos 265°25'39'' \\
&= 4977.92 \text{ ft}
\end{aligned}$$

The tangent of angle θ is found from

$$\begin{aligned}
\tan \theta &= \frac{x_P - x_M}{y_P - y_M} \\
&= \frac{1400.00 \text{ ft} - 1723.90 \text{ ft}}{5500.00 \text{ ft} - 4977.92 \text{ ft}} \\
&= -0.620403
\end{aligned}$$

Since θ is an angle and not an azimuth, the sign of $\tan \theta$ can be taken as positive. Therefore,

$$\begin{aligned}
\theta &= \tan^{-1} 0.620403 \\
&= 31°48'56''
\end{aligned}$$

Therefore, bearing PM is S $31°48'56''$ E.

32 *The answer is (B).*

The horizontal distance, HD, is given by

$$HD = SD \cos \beta$$

Therefore,

$$\begin{aligned}
HD_{true} &= (257.56 \text{ ft}) \cos 3°00'00'' \\
&= (257.56 \text{ ft})(0.998630) \\
&= 257.21 \text{ ft} \\
H_{incorrect} &= (257.56 \text{ ft}) \cos 4°00'00'' \\
&= (257.56 \text{ ft})(0.997564) \\
&= 256.93 \text{ ft}
\end{aligned}$$

The error is determined from the difference in these values.

$$\begin{aligned}
HD_{error} &= H_{true} - H_{incorrect} \\
&= 257.21 \text{ ft} - 256.93 \text{ ft} \\
&= 0.28 \text{ ft}
\end{aligned}$$

33 *The answer is (B).*

First determine the correction for sag of the tape.

$$\begin{aligned}
C_s &= \frac{w^2 L_s^3}{24 P_l^2} \\
&= \frac{\left(0.02 \dfrac{\text{lbf}}{\text{ft}}\right)^2 (100 \text{ ft})^3}{(24)(18 \text{ lbf})^2} \\
&= 0.05 \text{ ft}
\end{aligned}$$

The horizontal distance can now be determined.

$$HD = (D - C_s) \cos \beta$$
$$= (100.00 \text{ ft} - 0.05 \text{ ft}) \cos 1°40'00''$$
$$= 99.91 \text{ ft}$$

34   *The answer is (B).*

Since the tape is not level, the distance measured is a slope distance. The horizontal distance is therefore simple to find from the Pythagorean theorem.

$$HD = \sqrt{SD^2 - e^2}$$
$$= \sqrt{(39.56 \text{ ft})^2 - (2 \text{ ft})^2}$$
$$= 39.51 \text{ ft}$$

35   *The answer is (B).*

Clearly, measurement number 4 has a 1' blunder and should be discarded. Compute the mean of the remaining four angles, α_{mean}.

$$\alpha_{mean} = \frac{\Sigma \alpha}{4}$$
$$= \frac{\left(\begin{array}{c} 156°45'34'' + 156°45'38'' \\ + 156°45'29'' + 156°45'32'' \end{array}\right)}{4}$$
$$= 156°45'33.2''$$

Subtract the mean from each of the four angles, yielding four residuals, v_i.

$$+0.8'' \quad +4.8'' \quad -4.2'' \quad -1.2''$$

The standard error of a measurement, σ, is

$$\sigma = \sqrt{\frac{v_1^2 + v_2^2 + v_3^2 + v_4^2}{n-1}}$$
$$= \sqrt{\frac{(0.8'')^2 + (4.8'')^2 + (4.2'')^2 + (1.2'')^2}{3}}$$
$$= 3.8'' \quad (4'')$$

36   *The answer is (C).*

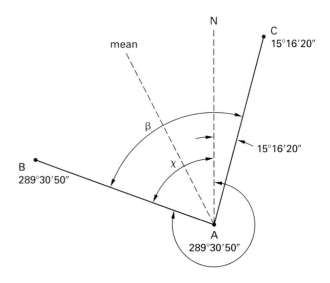

The angle between AB and north is

$$\chi = 360°00'00'' - 289°30'50''$$
$$= 70°29'10''$$

The angle between AB and AC is

$$\beta = 70°29'10'' + 15°16'20''$$
$$= 85°45'30''$$

The mean will occur at

$$\frac{\beta}{2} = \frac{85°45'30''}{2}$$
$$= 42°52'45''$$

Therefore,

$$Az_{mean \ line} = Az_{AB} + \frac{\beta}{2}$$
$$= 289°30'50'' + 42°52'45''$$
$$= 332°23'35''$$
$$\begin{array}{c} \text{bearing of} \\ \text{mean line} \end{array} = N \ (360°00'00 - 332°23'35'') \ W$$
$$= N \ 27°36'25'' \ W$$

37   *The answer is (C).*

The two-peg test will calibrate the deviation of the line of sight of the level from a level line. The elevation difference between P and Q can be determined twice: first from the backsight/foresight reading when set up at the midway point, and then from the two rod readings taken at the second setup. If these two elevation differences are not equal, the line of sight of the level will be in error.

38 *The answer is (A).*

The sum of the foresights minus the sum of the backsights is a check normally applied in differential leveling. It is equal to the difference in elevation between the starting point and the ending point. Since these are the same point in a loop, the answer is zero.

39 *The answer is (D).*

Differential leveling is usually adjusted as a linear function of distance. It is true that leveling misclosures are quite small, but they are always adjusted back through the intermediate points.

40 *The answer is (A).*

The change in horizontal distance PR is

$$\Delta \text{elev}_{\text{PR}} = \text{BS} - \text{FS}$$
$$= 4.67 \text{ ft} - 7.22 \text{ ft}$$
$$= -2.55 \text{ ft}$$

Use this to find the grade of course PR.

$$g_{\text{PR}} = \left(\frac{\Delta \text{elev}_{\text{PR}}}{D_{\text{PR}}}\right) \times 100\%$$
$$= \left(\frac{-2.55 \text{ ft}}{50 \text{ ft}}\right) \times 100\%$$
$$= -5.1\%$$

41 *The answer is (B).*

Use the temperature correction to determine the correction to the length, C.

$$C = L\alpha(T - T_f)$$
$$= (350.00 \text{ ft})\left(6.45 \times 10^{-6} \frac{\text{ft}}{\text{ft-}°\text{F}}\right)(73°\text{F} - 37°\text{F})$$
$$= 0.08 \text{ ft}$$

42 *The answer is (B).*

The correction to the length, C, is calculated as

$$C = \frac{(P - P_0)L}{aE}$$
$$= \frac{(35 \text{ lbf} - 15 \text{ lbf})(99.51 \text{ ft})}{(0.003 \text{ in}^2)\left(3.0 \times 10^7 \frac{\text{lbf}}{\text{in}^2}\right)}$$
$$= 0.02 \text{ ft}$$

Since the tape is elongated (stretched) with a 35 lbf pull, the reading on the tape will be 0.02 ft less than when the pull is 15 lbf.

The correct distance is

$$D = L + C = 99.51 \text{ ft} + 0.02 \text{ ft}$$
$$= 99.53 \text{ ft}$$

43 *The answer is (D).*

First compute the distance PT from coordinates.

$$D_{\text{PT}} = \sqrt{(x_{\text{P}} - x_{\text{T}})^2 + (y_{\text{P}} - y_{\text{T}})^2}$$
$$= \sqrt{\begin{array}{c}(500.67 \text{ ft} - 710.06 \text{ ft})^2 \\ + (876.45 \text{ ft} - 1005.34 \text{ ft})^2\end{array}}$$
$$= 245.88 \text{ ft}$$

The elevation of the tower can now be determined.

$$\text{elev}_{\text{T}} = \text{elev}_{\text{P}} + \text{HI} + D_{\text{PT}} \tan \text{VA}$$
$$= 1400.67 \text{ ft} + 5.04 \text{ ft} + (245.88 \text{ ft}) \tan 15°45'50''$$
$$= 1475.12 \text{ ft}$$

Calculations

44 *The answer is (B).*

First determine the angle R.

$$\text{angle R} = 180 - (\text{angle P} + \text{angle Q})$$
$$= 180° - (58°14'10'' + 87°34'00'')$$
$$= 34°11'50''$$

From the law of sines,

$$\frac{QR}{\sin P} = \frac{PQ}{\sin R}$$

Rearranging to solve for QR,

$$
\begin{aligned}
QR &= \frac{PQ \sin P}{\sin R} \\
&= \frac{(174.98 \text{ ft}) \sin 58°14'10''}{\sin 34°11'50''} \\
&= 264.70 \text{ ft}
\end{aligned}
$$

45 *The answer is (C).*

From the law of cosines,

$$b^2 = a^2 + c^2 - 2ac \cos B$$

Rearranging to solve for the cosine of B,

$$
\begin{aligned}
\cos B &= \frac{a^2 + c^2 - b^2}{2ac} \\
&= \frac{(330.56 \text{ ft})^2 + (380.02 \text{ ft})^2 - (210.90 \text{ ft})^2}{(2)(330.56 \text{ ft})(380.02 \text{ ft})} \\
&= 0.832699
\end{aligned}
$$

Therefore,

$$
\begin{aligned}
\text{angle B} &= \cos^{-1} 0.832699 \\
&= 33°37'23''
\end{aligned}
$$

46 *The answer is (D).*

Determine the elevation at point B.

$$
\begin{aligned}
\text{elev}_B &= \text{elev}_A - \Delta\text{elev}_{A-B} \\
&= 500.00 \text{ ft} - 19.65 \text{ ft} \\
&= 480.35 \text{ ft}
\end{aligned}
$$

The elevations of the top of the pole measured from points A and B are

$$
\begin{aligned}
\text{elev}_P \text{ from B} &= \text{HD}_{BP} \tan \beta_B + \text{elev}_B \\
&= \text{HD}_{BP} \tan 42°20'00'' + 480.35 \text{ ft} \\
\text{elev}_P \text{ from A} &= (\text{HD}_{BP} + \text{HD}_{AB})\tan \beta_A + \text{elev}_A \\
&= (\text{HD}_{BP} + 321.00 \text{ ft}) \tan 22°30'00'' \\
&\quad + 500.00 \text{ ft}
\end{aligned}
$$

Equating these two elevations,

$$
\begin{aligned}
\text{HD}_{BP} \tan 42°20'00'' &+ 480.35 \text{ ft} \\
&= (\text{HD}_{BP} + 321.00 \text{ ft}) \tan 22°30'00'' + 500.00 \text{ ft}
\end{aligned}
$$

Solving for HD_{BP},

$$
\begin{aligned}
\text{HD}_{BP}(0.910994 - 0.414214) &= 132.96 \text{ ft} + 500.00 \text{ ft} \\
&\quad - 480.35 \text{ ft} \\
\text{HD}_{BP} &= 307.20 \text{ ft}
\end{aligned}
$$

The elevation of the top of the pole above sea level is

$$
\begin{aligned}
\text{elev}_P &= \text{elev}_B + \text{HD}_{BP} \tan \beta_B \\
&= 480.35 \text{ ft} + \text{HD}_{BP} \tan 42°20'00'' \\
&= 480.35 \text{ ft} + (307.20 \text{ ft})(0.910994) \\
&= 760.21 \text{ ft}
\end{aligned}
$$

47 *The answer is (C).*

Since one bearing is west of north and the other is east of north,

$$
\begin{aligned}
\text{angle L} &= \text{bearing angle of LM} \\
&\quad + \text{bearing angle of LN} \\
&= 13°20'40'' + 39°59'10'' \\
&= 53°19'50''
\end{aligned}
$$

The area of the triangle, A, is

$$
\begin{aligned}
A &= \left(\frac{mn}{2}\right) \sin L \\
&= \left(\frac{(1298 \text{ ft})(1487 \text{ ft})}{2}\right) \sin 53°19'50'' \\
&= 774{,}071 \text{ ft}^2 \quad (774{,}100 \text{ ft}^2)
\end{aligned}
$$

48 *The answer is (B).*

First determine the radius, r.

$$
\begin{aligned}
r &= \frac{(100)(180)}{D\pi} \\
&= \frac{(100 \text{ ft})(180°)}{(2.5°)\pi} \\
&= 2291.83 \text{ ft}
\end{aligned}
$$

Solving for tangent distance, T,

$$T = r \tan \frac{I}{2} = (2291.83 \text{ ft}) \tan \frac{12°}{2}$$
$$= 240.88 \text{ ft} \quad (\text{sta } 2+40.88)$$

Therefore,

$$\text{sta}_{BC} = \text{sta}_{PI} - T$$
$$= (\text{sta } 23+44.78) - (\text{sta } 2+40.88)$$
$$= \text{sta } 21+03.90$$

49 *The answer is (B).*

First determine the total length of curve, L.

$$L_{\text{total}} = \left(\frac{I}{D}\right) L_{\text{between points}}$$
$$= \left(\frac{12°}{2.5°}\right)(100 \text{ ft})$$
$$= 480 \text{ ft} \quad (4.8000 \text{ sta})$$
$$\text{sta}_{EC} = \text{sta}_{BC} + L_{\text{total}}$$
$$= 21.0390 \text{ sta} + 4.8000 \text{ sta}$$
$$= 25.8390 \text{ sta}$$

The stationing of BC is sta 21+03.90, and the first point on the curve is sta 22+00. The second point on the curve is therefore sta 23+00.

$$L_{\text{point 2 to EC}} = \text{sta}_{EC} - \text{sta}_{\text{point 2}}$$
$$= 25.8390 \text{ sta} - 23.0000 \text{ sta}$$
$$= 2.8390 \text{ sta} \quad (283.90 \text{ ft})$$

The central angle generated by a 283.90 ft length of curve is

$$\Delta = D L_{\text{point 2 to EC}} = (2.5°)(2.8390 \text{ sta})$$
$$= 7°05'51''$$

Therefore, the deflection angle at the end of curve is

$$I_{EC} = \frac{\Delta}{2}$$
$$= \frac{7°05'51''}{2}$$
$$= 3°32'56''$$

50 *The answer is (C).*

The basic equation of a parabolic vertical curve is

$$\text{elev} = \text{elev}_{BVC} + g_1 x + \left(\frac{r}{2}\right) x^2$$

For the beginning of vertical curve (BVC),

$$\text{elev}_{BVC} = \text{elev}_{PVI} - g_1 \left(\frac{L}{2}\right)$$
$$= 310.50 \text{ ft} - (-1.5\%)\left(\frac{7 \text{ sta}}{2}\right)$$
$$= 315.75 \text{ ft}$$

The rate of change of grade is

$$r = \frac{g_2 - g_1}{L}$$
$$= \frac{2.0\% - (-1.5\%)}{7 \text{ sta}}$$
$$= 0.5\%/\text{sta}$$

The station of the beginning of vertical curve is

$$\text{sta}_{BVC} = \text{sta}_{PVI} - \frac{L}{2}$$
$$= 8.00 \text{ sta} - \frac{7.00 \text{ sta}}{2}$$
$$= 4.5 \text{ sta} \quad (\text{sta } 4+50)$$

Distance BVC to sta 7+10 in units of 100 ft is

$$D = \frac{\text{sta}_{7+10} - \text{sta}_{BVC}}{1000}$$
$$= \frac{710 \text{ ft} - 450 \text{ ft}}{100 \text{ ft}}$$
$$= 2.60$$

The elevation at sta 7+10, y, is

$$y = 315.75 \text{ ft} + (-1.5\%)(2.60)$$
$$+ \left(\frac{0.5 \dfrac{\%}{\text{sta}}}{2}\right)(2.60)^2$$
$$= 313.54 \text{ ft}$$

51   *The answer is (A).*

First compute the rate of change of grade, r.

$$r = \frac{g_2 - g_1}{L}$$
$$= \frac{2.0\% - (-1.5\%)}{7 \text{ sta}}$$
$$= 0.5\%/\text{sta}$$

The x value of the lowest point on the curve is

$$x = -\frac{g_1}{r}$$
$$= -\frac{-1.5\%}{0.5\dfrac{\%}{\text{sta}}}$$
$$= 3 \text{ sta from BVC}$$

The elevation of BVC is

$$\text{elev}_{\text{BVC}} = \text{elev}_{\text{PVI}} - g_1 \left(\frac{L}{2}\right)$$
$$= 310.50 \text{ ft} - (-1.5\%)\left(\frac{7 \text{ sta}}{2}\right)\left(100 \frac{\text{ft}}{\text{sta}}\right)$$
$$= 315.75 \text{ ft}$$

The elevation of the lowest point, P, is

$$\text{elev}_{\text{P}} = \text{elev}_{\text{BVC}} + g_1 x + \left(\frac{r}{2}\right)x^2$$
$$= 315.75 \text{ ft} + (-1.5\%)(3 \text{ sta})$$
$$+ \left(\frac{0.5\dfrac{\%}{\text{sta}}}{2}\right)(3 \text{ sta})^2$$
$$= 313.50 \text{ ft}$$

52   *The answer is (C).*

The elevation of the top of the rebar is

$$\text{elev}_{\text{top of rebar}} = \text{elev}_{\text{BM}} + \text{BS} - \text{FS}$$
$$= 390.66 \text{ ft} + 7.56 \text{ ft} - 1.01 \text{ ft}$$
$$= 397.21 \text{ ft}$$

Since the pipe surface is 1.5 ft below the top of the rebar,

$$\text{elev}_{\text{top of pipe}} = \text{elev}_{\text{top of rebar}} - L_{\text{rebar}}$$
$$= 397.21 \text{ ft} - (18 \text{ in})\left(\frac{1 \text{ ft}}{12 \text{ in}}\right)$$
$$= 395.71 \text{ ft}$$

53   *The answer is (D).*

Determine the height of instrument, HI, between points A and B.

$$\text{HI}_{\text{AB}} = \text{elev}_{\text{A}} + \text{BS}_{\text{A}}$$
$$= 1000.00 \text{ ft} + 3.34 \text{ ft}$$
$$= 1003.34 \text{ ft}$$

The height of instrument between points B and C can now be found.

$$\text{HI}_{\text{BC}} = \text{HI}_{\text{AB}} - \text{FS}_{\text{B}} + \text{BS}_{\text{B}}$$
$$= 1003.34 \text{ ft} - 2.89 \text{ ft} + 5.00 \text{ ft}$$
$$= 1005.45 \text{ ft}$$

54   *The answer is (A).*

The height of instrument, HI, at point D is

$$\text{HI}_{\text{D}} = \text{HI}_{\text{A}} + \sum \text{BS} - \sum \text{FS}$$
$$= 1000.00 \text{ ft} + (3.34 \text{ ft} + 5.00 \text{ ft} + 4.78 \text{ ft})$$
$$- (2.89 \text{ ft} + 3.04 \text{ ft} + 1.11 \text{ ft})$$
$$= 1006.08 \text{ ft}$$

Determine the misclosure.

$$e_H = \text{field elev}_{\text{D}} - \text{actual elev}_{\text{D}}$$
$$= 1006.08 \text{ ft} - 1006.18 \text{ ft}$$
$$= -0.10 \text{ ft}$$

55   *The answer is (B).*

The distance, D, in feet is

$$D = (2.86 \text{ mi})\left(5280 \frac{\text{ft}}{\text{mi}}\right)$$
$$= 15,100.80 \text{ ft}$$

Substitute into the formulas for earth curvature correction and refraction correction.

$$C_{\text{earth curvature}} = (0.0239)\left(\frac{D}{1000}\right)^2$$
$$= (0.0239)\left(\frac{15{,}100.80 \text{ ft}}{1000}\right)^2$$
$$= 5.45 \text{ ft}$$
$$C_{\text{refraction}} = -(0.0033)\left(\frac{D}{1000}\right)^2$$
$$= -(0.0033)\left(\frac{15{,}100.80 \text{ ft}}{1000}\right)^2$$
$$= -0.752 \text{ ft}$$

The combined correction can now be found.

$$C_{\text{total}} = C_{\text{earth curvature}} - C_{\text{refraction}}$$
$$= 5.45 \text{ ft} - 0.75 \text{ ft}$$
$$= 4.70 \text{ ft}$$

56   *The answer is (A).*

The measured line is too short to be affected by earth curvature and refraction. Therefore the elevation of point P can be found from the formula

$$\text{elev}_{\text{P}} = \text{elev}_{\text{T}} + \text{HI} + \text{H}_{\text{TP}} \tan \beta - \text{RR}$$

The height of instrument, HI, in feet is 4.58 ft.

Substituting,

$$\text{elev}_{\text{P}} = 760.00 \text{ ft} + 4.58 \text{ ft}$$
$$\qquad + (320.76 \tan -7°15'50'') - 3.99 \text{ ft}$$
$$= 719.71 \text{ ft}$$

57   *The answer is (D).*

The height of instrument, HI, is

$$\text{HI} = \text{elev}_{\text{BM}} + \text{BS}$$
$$= 2800.20 \text{ ft} + 2.55 \text{ ft}$$
$$= 2802.75 \text{ ft}$$

The elevation of the lath mark plus the foresight, FS, is equal to the height of instrument.

$$\text{HI} = \text{elev}_{\text{lath mark}} + \text{FS}$$

Rearranging to solve for the foresight, which is equal to the rod reading,

$$\text{RR} = \text{FS} = \text{HI} - \text{elev}_{\text{lath mark}}$$
$$= 2802.75 \text{ ft} - 2798.11 \text{ ft}$$
$$= 4.64 \text{ ft}$$

58   *The answer is (B).*

Compute the (x, y) coordinates for Q and R, and then compute distance PR from the coordinate differences.

$$y_{\text{Q}} = y_{\text{P}} + D_{\text{PQ}} \cos \text{Az}_{\text{PQ}}$$
$$= 1000.00 \text{ ft} + (151.84 \text{ ft}) \cos 43°16'00''$$
$$= 1110.57 \text{ ft}$$
$$x_{\text{Q}} = x_{\text{P}} + D_{\text{PQ}} \sin \text{Az}_{\text{PQ}}$$
$$= 500.00 \text{ ft} + (151.84 \text{ ft}) \sin 43°16'00''$$
$$= 604.07 \text{ ft}$$

Compute the (x, y) coordinates for R in the same manner.

$$y_{\text{R}} = y_{\text{Q}} + D_{\text{QR}} \cos \text{Az}_{\text{QR}}$$
$$= 1110.57 \text{ ft} + (145.22 \text{ ft}) \cos 111°29'00''$$
$$= 1057.39 \text{ ft}$$
$$x_{\text{R}} = x_{\text{Q}} + D_{\text{QR}} \sin \text{Az}_{\text{QR}}$$
$$= 604.07 \text{ ft} + (145.22 \text{ ft}) \sin 111°29'00''$$
$$= 739.20 \text{ ft}$$

Distance PR is

$$D_{\text{PR}} = \sqrt{(y_{\text{P}} - y_{\text{R}})^2 + (x_{\text{P}} - x_{\text{R}})^2}$$
$$= \sqrt{\begin{array}{l}(1000.00 \text{ ft} - 1057.39 \text{ ft})^2 \\ \quad + (500.00 \text{ ft} - 739.20 \text{ ft})^2\end{array}}$$
$$= 245.99 \text{ ft}$$

59   *The answer is (B).*

The fractional misclosure is given by $1/n$, where n is the total traverse length divided by the vector misclosure.

The vector misclosure is

$$e_C = \sqrt{e_x^2 + e_y^2}$$
$$= \sqrt{(0.54 \text{ ft})^2 + (0.23 \text{ ft})^2}$$
$$= 0.587 \text{ ft}$$

The total traverse length is

$$L_{\text{total}} = \sum L$$
$$= 1367.2 \text{ ft} + 2756.0 \text{ ft} + 1163.8 \text{ ft}$$
$$\quad + 998.3 \text{ ft} + 3672.5 \text{ ft}$$
$$= 9957.8 \text{ ft}$$

Therefore,

$$n = \frac{L_{\text{total}}}{e_C} = \frac{9957.8 \text{ ft}}{0.587 \text{ ft}} = 16{,}964$$

The fractional misclosure is most nearly 1/17,000.

60   *The answer is (B).*

Determine the total length of the traverse, in feet.

$$L_{\text{perimeter}} = (2)(1.6 \text{ mi} + 2.4 \text{ mi}) \left(5280 \, \frac{\text{ft}}{\text{mi}} \right)$$
$$= 42{,}240 \text{ ft}$$

The allowable misclosure is

$$e_{H,\text{allowable}} = \frac{L_{\text{perimeter}}}{80{,}000} = \frac{42{,}240 \text{ ft}}{80{,}000}$$
$$= 0.53 \text{ ft}$$

61   *The answer is (D).*

This is a bearing/bearing intersection requiring the azimuths BP and AP. If the azimuths are denoted by Az, the coordinates of P may be computed from

$$\text{Az}_{\text{BP}} = 360° - 30°$$
$$= 330°$$
$$\text{Az}_{\text{PB}} = \text{Az}_{\text{BP}} - 180°$$
$$= 330° - 180°$$
$$= 150°$$
$$\text{Az}_{\text{PA}} = \text{Az}_{\text{PB}} + \alpha$$
$$= 150° + 92°10'00''$$
$$= 242°10'00''$$
$$\text{Az}_{\text{AP}} = \text{Az}_{\text{PA}} - 180°$$
$$= 62°10'00''$$

Now find the (x, y) coordinates.

$$y_P = \frac{(x_A - x_B) - y_A \tan \text{Az}_{\text{AP}} + y_B \tan \text{Az}_{\text{BP}}}{\tan \alpha_B - \tan \alpha_A}$$
$$= \frac{\begin{array}{c}(500 \text{ ft} - 1000 \text{ ft}) - (1100 \text{ ft}) \tan 62°10'00'' \\ + (1000 \text{ ft}) \tan 330°\end{array}}{\tan 330° - \tan 62°10'00''}$$
$$= 1278.96 \text{ ft}$$
$$x_P = (y_P - y_A) \tan \text{Az}_{\text{AP}} + x_A$$
$$= (1278.96 \text{ ft} - 1100 \text{ ft}) \tan 62°10'00'' + 500 \text{ ft}$$
$$= 838.94 \text{ ft}$$

62   *The answer is (B).*

First compute distance CP.

$$D_{\text{CP}} = \sqrt{(x_P - x_C)^2 + (y_P - y_C)^2}$$
$$= \sqrt{(4400 \text{ ft} - 5000 \text{ ft})^2 + (7700 \text{ ft} - 5000 \text{ ft})^2}$$
$$= 2765.86 \text{ ft}$$

Now compute angle C.

$$\cos C = \frac{2200 \text{ ft}}{D_{\text{CP}}}$$
$$= \frac{2200 \text{ ft}}{2765.86 \text{ ft}}$$
$$= 0.795412$$

Therefore,

$$\text{angle } C = \cos^{-1} 0.795412$$
$$= 37°18'21''$$

The tangent distance can now be found.

$$T = r \tan C$$
$$= (2200 \text{ ft}) \tan 37°18'21''$$
$$= 1676.30 \text{ ft}$$

63   *The answer is (C).*

The geoid is the surface of mean sea level extended over the entire globe or the surface of the sea if small channels were cut through all the mountains, allowing the sea to flow through all land masses. The geodetic reference surface mentioned in option (A) is called the spheroid or ellipsoid. The height of mean sea level above or below the spheroid is called the geoidal undulation.

64 *The answer is (B).*

The ditch end area can be determined from

$$A_{\text{end}} = \frac{bD}{2}$$

Therefore,

$$A_{\text{end 1}} = \frac{(30 \text{ ft})(15.9 \text{ ft})}{2} = 238.50 \text{ ft}^2$$

$$A_{\text{end 2}} = \frac{(30 \text{ ft})(18.94 \text{ ft})}{2} = 284.10 \text{ ft}^2$$

It follows that the volume of earth removed from the ditch is

$$\begin{aligned} V_e &= \left(\frac{A_1 + A_2}{2}\right) L \\ &= \left(\frac{238.50 \text{ ft}^2 + 284.10 \text{ ft}^2}{2}\right)(120 \text{ ft})\left(\frac{1 \text{ yd}^3}{27 \text{ ft}^3}\right) \\ &= 1161.3 \text{ yd}^3 \quad (1160 \text{ yd}^3) \end{aligned}$$

65 *The answer is (C).*

The volume of dirt excavated is

$$V = \pi r^2 D = \frac{\pi(12.5 \text{ ft})^2(10.5 \text{ ft})}{27 \dfrac{\text{ft}^3}{\text{yd}^3}}$$

$$= 190.90 \text{ yd}^3$$

Calculate the total excavation and overhaul costs.

$$\begin{aligned} C_{\text{excavation,total}} &= V C_{\text{excavation}} \\ &= (190.90 \text{ yd}^3)\left(2.50 \dfrac{\$}{\text{yd}^3}\right) \\ &= \$477.24 \\ C_{\text{overhaul,total}} &= V C_{\text{overhaul}}(\text{distance} - \text{freehaul}) \\ &= (190.90 \text{ yd}^3)\left(0.60 \dfrac{\dfrac{\$}{\text{yd}^3}}{\text{sta}}\right) \\ &\quad \times (8 \text{ sta} - 4 \text{ sta}) \\ &= \$458.16 \end{aligned}$$

The total cost of the move is then

$$\begin{aligned} C_{\text{total}} &= C_{\text{excavation,total}} + C_{\text{overhaul,total}} \\ &= \$477.24 + \$458.16 \\ &= \$935.40 \quad (\$930) \end{aligned}$$

66 *The answer is (D).*

The reference system for NAD 83 (the North American Datum of 1983) is the Geodetic Reference System of 1980, or the GRS 80. The Clarke spheroid is the basis of the NAD 1927 State Plane System.

67 *The answer is (D).*

In the NAD 1927 state plane system there are seven zones for California. For the NAD 83 state plane system, zone 7 (covering Los Angeles) was incorporated into zone 5, resulting in six zones for NAD 83.

68 *The answer is (C).*

The first three points on the curve will have stations of sta 18+50, sta 19+00, and sta 19+50. The tangent offset is thus required for sta 19+50.

Start by determining the arc length.

$$\begin{aligned} l &= \text{sta}_{\text{point 3}} - \text{sta}_{\text{PC}} \\ &= 19.50 \text{ sta} - 18.1899 \text{ sta} \\ &= 1.3101 \text{ sta} \quad (131.01 \text{ ft}) \end{aligned}$$

The central angle, Δ, can now be found.

$$\begin{aligned} \Delta &= \frac{l(360)}{2\pi r} \\ &= \frac{(131.01 \text{ ft})(360°)}{2\pi(2400 \text{ ft})} \\ &= 3°07'39'' \end{aligned}$$

The tangent offset, x, is

$$\begin{aligned} x &= r(1 - \cos\Delta) \\ &= (2400 \text{ ft})(1 - \cos 3°07'39'') \\ &= 3.57 \text{ ft} \end{aligned}$$

69 *The answer is (A).*

Compute the latitude and departure differences, d_y and d_x, respectively.

$$d_x = D \sin \text{Az}$$
$$d_y = D \cos \text{Az}$$

Compute the (x, y) coordinates of B and C.

$$
\begin{aligned}
y_\text{B} &= y_\text{A} + d_{y,\text{AB}} \\
&= 1000.00 \text{ ft} + (267.56 \text{ ft}) \cos 319°54'20'' \\
&= 1204.68 \text{ ft} \\
x_\text{B} &= x_\text{A} + d_{x,\text{AB}} \\
&= 500.00 \text{ ft} + (267.56 \text{ ft}) \sin 319°54'20'' \\
&= 327.68 \text{ ft} \\
y_\text{C} &= y_\text{B} + d_{y,\text{BC}} \\
&= 1204.68 \text{ ft} + (189.55 \text{ ft}) \cos 18°44'30'' \\
&= 1384.18 \text{ ft} \\
x_\text{C} &= x_\text{B} + d_{x,\text{BC}} \\
&= 327.68 \text{ ft} + (189.55 \text{ ft}) \sin 18°44'30'' \\
&= 388.58 \text{ ft}
\end{aligned}
$$

The misclosures of x and y are then

$$
\begin{aligned}
e_y &= \text{computed } y \text{ coordinate} - \text{known } y \text{ coordinate} \\
&= 1384.18 \text{ ft} - 1384.38 \text{ ft} \\
&= -0.20 \text{ ft} \\
e_x &= \text{computed } x \text{ coordinate} - \text{known } x \text{ coordinate} \\
&= 388.58 \text{ ft} - 388.28 \text{ ft} \\
&= +0.30 \text{ ft}
\end{aligned}
$$

70 *The answer is (A).*

The elevation of point BMb is

$$
\begin{aligned}
\text{elev}_\text{BMb} &= \text{elev}_\text{BMa} + \text{BS}_1 - \text{FS}_1 + \text{BS}_2 \\
&\quad - \text{FS}_2 + \ldots + \text{BS}_n - \text{FS}_n \\
&= 783.67 \text{ ft} + 4.67 \text{ ft} - 5.59 \text{ ft} \\
&\quad + 3.0 \text{ ft} - 6.51 \text{ ft} + 4.26 \text{ ft} \\
&\quad - 5.74 \text{ ft} + 3.22 \text{ ft} - 7.38 \text{ ft} \\
&= 773.60 \text{ ft}
\end{aligned}
$$

The change in elevation from BMb to BMc is

$$
\begin{aligned}
\Delta\text{elev}_\text{BMb-BMc} &= \text{elev}_\text{BMc} - \text{elev}_\text{BMb} \\
&= 763.60 \text{ ft} - 773.60 \text{ ft} \\
&= -10.00 \text{ ft}
\end{aligned}
$$

The grade of line BMb−BMc can now be found.

$$
\begin{aligned}
g_\text{BMb-BMc} &= \frac{\Delta\text{elev}_\text{BMb-BMc}}{\text{HD}_\text{BMb-BMc}} \\
&= \left(\frac{-10.00 \text{ ft}}{200.00 \text{ ft}}\right) \times 100\% \\
&= -5\%
\end{aligned}
$$

71 *The answer is (C).*

The elevation of point P can be found from

$$\text{elev}_\text{P} = \text{elev}_\text{T} + \text{HI} - \text{HD} \tan \alpha - \text{RR}$$

Rearranging,

$$
\begin{aligned}
\text{elev}_\text{T} &= \text{elev}_\text{P} - \text{HI} + \text{HD} \tan \alpha + \text{RR} \\
&= 1703.99 \text{ ft} - 5.02 \text{ ft} \\
&\quad + (220.85 \text{ ft}) \tan 7°10'10'' + 4.22 \text{ ft} \\
&= 1730.97 \text{ ft}
\end{aligned}
$$

72 *The answer is (B).*

Because of the inadequacies of the NAD 1927 national survey adjustment, a decision was made in the 1970s to recompute the national network. This new adjustment was commenced in 1974 and completed in 1986. It became known as the NAD 1983 adjustment. One outcome of this massive computation was the introduction of a new unit called the U.S. survey foot.

73 *The answer is (D).*

The deflection angle at CT, I, is half the central angle, Δ. It follows that

$$
\begin{aligned}
\tan I &= \frac{x}{T} = \frac{19.464 \text{ ft}}{278.346 \text{ ft}} \\
&= 0.069927 \\
I &= \tan^{-1} 0.069927 \\
&= 4° \\
\Delta &= 2I = (2)(4°) \\
&= 8°
\end{aligned}
$$

74 *The answer is (C).*

Use the following formula to determine the volume of section AB.

$$V = \left(\frac{A_A + A_B}{2} \right) S$$

Use the following formula to find the areas of the sections.

$$A = c \left(\frac{d_l + d_r}{2} \right) + b \left(\frac{h_l + h_r}{4} \right)$$

The areas of sections A and B are

$$A_A = (3.01 \text{ ft}) \left(\frac{38.91 \text{ ft} + 42.98 \text{ ft}}{2} \right)$$
$$+ (60 \text{ ft}) \left(\frac{5.81 \text{ ft} + 6.44 \text{ ft}}{4} \right)$$
$$= 306.99 \text{ ft}^2$$

$$A_B = (3.21 \text{ ft}) \left(\frac{37.00 \text{ ft} + 44.95 \text{ ft}}{2} \right)$$
$$+ (60 \text{ ft}) \left(\frac{5.76 \text{ ft} + 7.94 \text{ ft}}{4} \right)$$
$$= 337.03 \text{ ft}^2$$

Therefore,

$$V = \left(\frac{306.99 \text{ ft}^2 + 337.03 \text{ ft}^2}{2} \right) \left(\frac{100 \text{ ft}}{27 \frac{\text{ft}^3}{\text{yd}^3}} \right)$$
$$= 1192.64 \text{ yd}^3 \quad (1190 \text{ yd}^3)$$

75 *The answer is (C).*

By far, the largest source of error in trigonometric leveling is refraction. On longer lines, refraction has a significant effect on the elevation angle, and therefore on the determination of the elevation of the point sighted.

76 *The answer is (C).*

The elevation of point Q is found from

$$\text{elev}_Q = \text{elev}_P + \left(\sum \text{BS} - \sum \text{FS} \right)$$
$$= 1000.00 \text{ ft}$$
$$+ (4.68 \text{ ft} + 3.99 \text{ ft} + 2.07 \text{ ft} + 5.55 \text{ ft})$$
$$- (2.12 \text{ ft} + 1.63 \text{ ft} + 1.64 \text{ ft} + 4.71 \text{ ft})$$
$$= 1006.19 \text{ ft}$$

77 *The answer is (A).*

Compute the latitude difference, d_y, for courses AB and BC.

$$d_y = D \cos \text{Az}$$

Therefore,

$$d_{y,AB} = (267.45 \text{ ft}) \cos 127°03'20''$$
$$= -161.162 \text{ ft}$$
$$d_{y,BC} = (180.99 \text{ ft}) \cos 190°30'30''$$
$$= -177.955 \text{ ft}$$

It follows that

$$y_C = y_A + d_{y,AB} + d_{y,BC}$$
$$= 850.00 \text{ ft} + (-161.162 \text{ ft}) + (-177.955 \text{ ft})$$
$$= 510.883 \text{ ft}$$

78 *The answer is (C).*

The offset can be found from

$$x = (r_1 + r_2)(1 - \cos I)$$
$$= (1500 \text{ ft} + 1200 \text{ ft})(1 - \cos 15°20'00'')$$
$$= 96.11 \text{ ft}$$

79 *The answer is (B).*

Use the following formula to determine the area.

$$A = \sqrt{s(s-a)(s-b)(s-c)}$$
$$s = \frac{a+b+c}{2}$$
$$= \frac{14.98 \text{ ch} + 15.03 \text{ ch} + 19.76 \text{ ch}}{2}$$
$$= 24.885 \text{ ch}$$

Therefore,

$$A = \sqrt{\begin{array}{c} (24.885 \text{ ch})(24.885 \text{ ch} - 14.98 \text{ ch}) \\ \times (24.885 \text{ ch} - 15.03 \text{ ch}) \\ \times (24.885 \text{ ch} - 19.76 \text{ ch}) \end{array}} \left(\frac{1 \text{ ac}}{10 \text{ ch}^2} \right)$$
$$= 11.158 \text{ ac} \quad (11.16 \text{ ac})$$

80 *The answer is (B).*

Using the law of cosines,

$$\cos A = \frac{(AC)^2 + (AB)^2 - (BC)^2}{2(AB)(AC)}$$
$$= \frac{(410)^2 + (450)^2 - (220)^2}{(2)(410)(450)}$$
$$= 0.873171$$

Therefore,

$$\text{angle A} = \cos^{-1} 0.873171$$
$$= 29°10'15''$$

81 *The answer is (B).*

The area of the triangle bounded by the radii and the 100 ft line is

$$A = \tfrac{1}{2}bh$$

Using the Pythagorean theorem, the height of the triangle is

$$h = \sqrt{r^2 - \left(\frac{L}{2}\right)^2} = \sqrt{(900 \text{ ft})^2 - \left(\frac{100 \text{ ft}}{2}\right)^2}$$
$$= 898.61 \text{ ft}$$

The area of the triangle is

$$A_t = \tfrac{1}{2}bh$$
$$= \left(\frac{1}{2}\right)(100 \text{ ft})(898.61 \text{ ft})$$
$$= 44{,}931 \text{ ft}^2$$

The area of the segment bounded by the two radii, A_s, is

$$A_s = \pi r^2 \left(\frac{\theta}{360}\right)$$

Determine the angle θ.

$$\theta = 2\sin^{-1}\left(\frac{\dfrac{L}{2}}{r}\right)$$
$$= 2\sin^{-1}\frac{\dfrac{100 \text{ ft}}{2}}{900 \text{ ft}}$$
$$= 6.3695°$$

Therefore,

$$A_s = \pi(900 \text{ ft})^2 \left(\frac{6.3695°}{360°}\right)$$
$$= 45{,}023 \text{ ft}^2$$

The area of the segment bounded by the curve and the 100 ft line is

$$A_s - A_t = 45{,}023 \text{ ft}^2 - 44{,}931 \text{ ft}^2$$
$$= 92 \text{ ft}^2$$

82 *The answer is (C).*

Determine the azimuth from

$$Az = 180 + \tan^{-1}\frac{\text{dep}}{\text{lat}}$$
$$= 180° + \tan^{-1}\frac{-200 \text{ ft}}{-100 \text{ ft}}$$
$$= 243°26'06''$$

83 *The answer is (A).*

Draw a diagram of the points in the (x, y) plane.

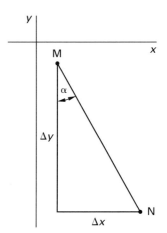

Angle α is found from

$$\alpha = \tan^{-1}\left|\frac{d_x}{d_y}\right|$$
$$= \tan^{-1}\left|\frac{500 - 100}{800 - 100}\right|$$
$$= \tan^{-1} 0.571429$$
$$= 29°44'42''$$

Therefore,

$$\begin{aligned}\text{bearing MN} &= \text{S } \alpha \text{ E}\\ &= \text{S } 29.7449° \text{ E}\\ &= \text{S } 29°44'42'' \text{ E}\end{aligned}$$

84 *The answer is (B).*

The area of triangle ABC is determined from the formula

$$A = \tfrac{1}{2}D_a D_b \sin C$$

First calculate angle C.

$$\text{angle C} = \text{Az}_{CA} - \text{Az}_{CB}$$

Azimuths CB and CA are

$$\begin{aligned}\text{Az}_{CB} &= 82°15'00''\\ \text{Az}_{CA} &= 180°00'00'' - 40°00'00''\\ &= 140°00'00''\end{aligned}$$

Therefore,

$$\begin{aligned}\text{angle C} &= 140°00'00'' - 82°15'00''\\ &= 57°45'00''\end{aligned}$$

The area of triangle ABC can now be found.

$$\begin{aligned}A &= \left(\tfrac{1}{2}\right)(486.99 \text{ ft})(432.12 \text{ ft})\sin 57°45'00''\\ &= 88{,}987 \text{ ft}^2 \quad (89{,}000 \text{ ft}^2)\end{aligned}$$

85 *The answer is (B).*

Convert angles to decimal degrees.

$$\begin{aligned}15°05' &= 15° + \left(\frac{5'}{\frac{60'}{1°}}\right)\\ &= 15.083°\\ 11°45' &= 11° + \left(\frac{45'}{\frac{60'}{1°}}\right)\\ &= 11.750°\end{aligned}$$

Next, convert angles to radians.

$$\begin{aligned}15.083° &= \frac{15.083°(\pi \text{ rad})}{180°}\\ &= 0.26325 \text{ rad}\\ 11.750° &= \frac{11.750°(\pi \text{ rad})}{180°}\\ &= 0.20508 \text{ rad}\end{aligned}$$

The length of curve is

$$\begin{aligned}L &= \Delta_1 r_1 + \Delta_2 r_2\\ &= (0.2633 \text{ rad})(2000 \text{ ft}) + (0.2051 \text{ rad})(2450 \text{ ft})\\ &= 1028.94 \text{ ft}\end{aligned}$$

86 *The answer is (A).*

The correction to departure AB is

$$\begin{aligned}C &= -\left(\frac{D_{AB}}{D_{\text{total}}}\right)e_E\\ &= -\left(\frac{201.66 \text{ ft}}{589.01 \text{ ft}}\right)(0.24 \text{ ft})\\ &= -0.08 \text{ ft}\end{aligned}$$

The balanced departure of line AB can now be determined.

$$\begin{aligned}\text{dep}_{AB,\text{balanced}} &= \text{dep}_{AB,\text{computed}} - C\\ &= 154.86 \text{ ft} - 0.08 \text{ ft}\\ &= 154.78 \text{ ft}\end{aligned}$$

87 *The answer is (B).*

The sea-level scale factor is

$$\begin{aligned}\text{sea-level scale factor} &= \frac{r}{r + \text{elev}}\\ &= \frac{20{,}906{,}000 \text{ ft}}{20{,}906{,}000 \text{ ft} + 5400 \text{ ft}}\\ &= 0.9997418\end{aligned}$$

Using this value to correct the distance measurement,

$$\begin{aligned}D_{\text{corrected}} &= (\text{sea-level scale factor})D\\ &= (0.9997418)(2856.11 \text{ ft})\\ &= 2855.37 \text{ ft}\end{aligned}$$

88 *The answer is (A).*

The dimensions of the midsections are the means of the dimensions of S_1 and S_2. For the midsection,

$$b_{\text{mean}} = \frac{S_{1,b} + S_{2,b}}{2} = \frac{40.0 \text{ ft} + 50.0 \text{ ft}}{2}$$
$$= 45.0 \text{ ft}$$

$$c_{\text{mean}} = \frac{S_{1,c} + S_{2,c}}{2} = \frac{5.67 \text{ ft} + 7.14 \text{ ft}}{2}$$
$$= 6.41 \text{ ft}$$

If A is the area of a cross section, then

$$A = c(b + sc)$$

For the midsection,

$$A = (6.41 \text{ ft})(45.0 \text{ ft} + (0.5)(6.41 \text{ ft}))$$
$$= 308.99 \text{ ft}^2$$

89 *The answer is (B).*

The station of the tangent point is

$$\text{sta}_{\text{tangent point}} = \text{sta}_{\text{PI}} - T$$
$$= 18.3790 \text{ sta} - 2.1009 \text{ sta}$$
$$= 16.2781 \text{ sta}$$

The station of the first point on the curve is therefore sta 17+00.00.

The distance from the tangent point to the first curve point is

$$D = 100(\text{sta}_{\text{first point}} - \text{sta}_{\text{tangent point}})$$
$$= (100)(17.0000 \text{ sta} - 16.2781)$$
$$= 72.19 \text{ ft}$$

90 *The answer is (B).*

The area of such a profile can be determined using the following equation.

$$A = 0.5(a(h_0 + h_1) + b(h_1 + h_2) + c(h_2 + h_3) + ...)$$

The parameters of the equation are illustrated as follows.

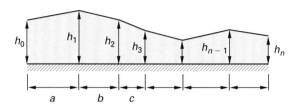

Applying this equation to the data given in the problem statement,

$$A = (0.5)((10.9 \text{ ft})(2.1 \text{ ft} + 3.5 \text{ ft})$$
$$+ (21.6 \text{ ft})(3.5 \text{ ft} + 2.4 \text{ ft})$$
$$+ (18.0 \text{ ft})(2.4 \text{ ft} + 3.3 \text{ ft}))$$
$$= 145.5 \text{ ft}^2 \quad (150 \text{ ft}^2)$$

91 *The answer is (C).*

The station of the beginning of vertical curve is

$$\text{sta}_{\text{BVC}} = \text{sta}_{\text{PVI}} - \frac{L}{2}$$
$$= \text{sta } 16+40.00 - \frac{700 \text{ ft}}{2}$$
$$= \text{sta } 12+90.00$$

Find the rate of change of grade.

$$r = \frac{g_2 - g_1}{L} = \frac{-3\% - 2\%}{7 \text{ sta}}$$
$$= -0.7143\%/\text{sta}$$

The x coordinate of the highest point is

$$x_{\text{HP}} = -\frac{g_1}{r} = -\frac{2\%}{-0.7143 \dfrac{\%}{\text{sta}}}$$
$$= 2.80 \text{ sta}$$

The station of the highest point is

$$\text{sta}_{\text{HP}} = \text{sta}_{\text{BVC}} + x_{\text{HP}}$$
$$= \text{sta } 12+90.00 + \text{sta } 2+80.00$$
$$= \text{sta } 15+70.00$$

92 *The answer is (C).*

First find the offset, x.

$$x = 0.5 x_{\text{total}}$$
$$= (0.5)(24.00 \text{ ft})$$
$$= 12.00 \text{ ft}$$

It is known that

$$x = r(1 - \cos I)$$

Therefore,

$$12.00 \text{ ft} = (1000.00 \text{ ft})(1 - \cos I)$$

Solving for the cosine of I,

$$\cos I = 1 - \frac{12.00 \text{ ft}}{1000.00 \text{ ft}}$$
$$= 0.988000$$

Therefore,

$$I = \cos^{-1} 0.988000 = 8°53'06''$$

It is known that

$$y = \frac{D}{2} = r \sin I$$
$$= (1000.00 \text{ ft}) \sin 8°53'06''$$
$$= 154.45 \text{ ft}$$

Solving for D yields

$$D = (2)(154.45 \text{ ft})$$
$$= 308.90 \text{ ft}$$

93 *The answer is (C).*

The station of the beginning of vertical curve is

$$\text{sta}_{\text{BVC}} = \text{sta}_{\text{PVI}} - \frac{L}{2}$$
$$= \text{sta } 28{+}15.00 - \frac{8 \text{ sta}}{2}$$
$$= \text{sta } 24{+}15.00$$

And the elevation of the beginning of vertical curve is

$$\text{elev}_{\text{BVC}} = \text{elev}_{\text{PVI}} - g_1 \left(\frac{L}{2} \right)$$
$$= 644.73 \text{ ft} - (-2\%)\left(\frac{8 \text{ sta}}{2} \right)$$
$$= 652.73 \text{ ft}$$

Find the rate of change of grade.

$$r = \frac{g_2 - g_1}{L} = \frac{1.5\% - (-2\%)}{8 \text{ sta}}$$
$$= 0.4375\%/\text{sta}$$

The first point, P, on the curve is the next 100 ft station past BVC, or sta 25+00.00.

$$D_{\text{BVC-P}} = \text{sta}_{\text{P}} - \text{sta}_{\text{BVC}}$$
$$= \text{sta } 25{+}00 - \text{sta } 24{+}15.00$$
$$= 85.00 \text{ ft}$$

The x coordinate of the first full station of the curve, P, is

$$x_{\text{P}} = \frac{D_{\text{BVC-P}}}{100} = \frac{85.00 \text{ ft}}{100 \dfrac{\text{ft}}{\text{sta}}}$$
$$= 0.85 \text{ sta}$$

So the elevation of P is

$$\text{elev}_{\text{P}} = \text{elev}_{\text{BVC}} + g_1 x + \left(\frac{r}{2} \right) x^2$$
$$= 652.73 \text{ ft} + (-2\%)(0.85 \text{ sta})$$
$$+ \left(\frac{0.4375 \dfrac{\%}{\text{sta}}}{2} \right)(0.85 \text{ sta})^2$$
$$= 651.19 \text{ ft}$$

94 *The answer is (D).*

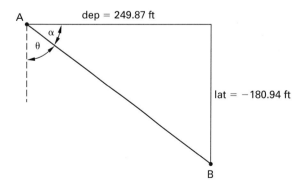

The tangent of angle α is

$$\tan \alpha = \left| \frac{\text{lat}}{\text{dep}} \right| = \left| \frac{-180.94 \text{ ft}}{249.87 \text{ ft}} \right|$$
$$= 0.724137$$

Therefore,

$$\alpha = \tan -1(0.724137) = 35°54'35''$$

Angle θ can now be found.

$$\theta = 90 - \alpha$$
$$= 90° - 35°54'35''$$
$$= 54°05'25''$$

Therefore,

$$\text{bearing AB} = \text{S } \theta \text{ E} = \text{S } 54°05'25'' \text{ E}$$

95 *The answer is (A).*

Start by summing the distances.

$$D_{\text{total}} = D_{\text{AB}} + D_{\text{BC}} + D_{\text{CD}} + D_{\text{DE}}$$
$$= 156.45 \text{ ft} + 145.89 \text{ ft} + 122.22 \text{ ft} + 167.89 \text{ ft}$$
$$= 592.45 \text{ ft}$$

Next determine the misclosures in N and E.

$$e_{\text{N}} = 185.18 \text{ ft} - 185.37 \text{ ft} = -0.19 \text{ ft}$$
$$e_{\text{E}} = 358.33 \text{ ft} - 358.00 \text{ ft} = 0.33 \text{ ft}$$

Find the corrections.

$$\text{N correction to B} = \left(\frac{D_{\text{AB}}}{D_{\text{total}}} \right) e_{\text{N}}$$
$$= \left(\frac{156.45 \text{ ft}}{592.45 \text{ ft}} \right)(-0.19 \text{ ft})$$
$$= -0.05 \text{ ft}$$
$$\text{E correction to B} = \left(\frac{D_{\text{AB}}}{D_{\text{total}}} \right) e_{\text{E}}$$
$$= \left(\frac{156.45 \text{ ft}}{592.45 \text{ ft}} \right)(0.33 \text{ ft})$$
$$= 0.09 \text{ ft}$$

The balanced coordinates are

$$\text{balanced N}_{\text{B}} = \text{N}_{\text{B}} - \text{N correction to B}$$
$$= 254.07 \text{ ft} - (-0.05 \text{ ft})$$
$$= 254.12 \text{ ft}$$
$$\text{balanced E}_{\text{B}} = \text{E}_{\text{B}} - \text{E correction to B}$$
$$= 127.17 \text{ ft} - 0.09 \text{ ft}$$
$$= 127.08 \text{ ft}$$

96 *The answer is (C).*

Find the radius of the circle.

$$r = \frac{c}{\pi D} = \frac{18,000 \text{ ft}}{\pi(2°)}$$
$$= 2864.79 \text{ ft}$$

It follows that the tangent distance is

$$T = \sqrt{E^2 - r^2}$$
$$= \sqrt{(4160.10 \text{ ft})^2 - (2864.79 \text{ ft})^2}$$
$$= 3016.52 \text{ ft}$$

97 *The answer is (A).*

First find the central angle for TC–O–P.

$$\Delta = \left(\frac{L}{100} \right) D$$
$$= \left(\frac{145.00 \text{ ft}}{100 \text{ ft}} \right)(2.5°)$$
$$= 3°37'30''$$

The tangential deflection angle for S–TC–P is

$$I = \frac{\Delta}{2}$$
$$= \frac{3°37'30''}{2}$$
$$= 1°48'45''$$

The bearings can now be found.

$$\begin{aligned}
\text{bearing TC–S} &= \text{bearing TC–Q} + 180 \\
&= \text{S } 66°14'40'' \text{ W} + 180° \\
&= \text{N } 66°14'40'' \text{ E} \\
\text{bearing TC–P} &= \text{bearing TC–S} + I \\
&= \text{N } 66°14'40'' \text{ E} + 1°48'45'' \\
&= \text{N } 68°03'25'' \text{ E}
\end{aligned}$$

98 *The answer is (C).*

Sum the latitudes and departures, respectively.

$$\begin{aligned}
\text{lat}_{\text{total}} &= \sum \text{lat} \\
&= 100 \text{ ft} - 30 \text{ ft} - 90 \text{ ft} \\
&= -20 \text{ ft} \\
\text{dep}_{\text{total}} &= \sum \text{dep} \\
&= 100 \text{ ft} + 90 \text{ ft} + 20 \text{ ft} \\
&= 210 \text{ ft}
\end{aligned}$$

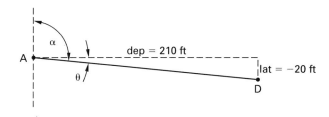

The tangent of angle θ is

$$\begin{aligned}
\tan \theta &= \left| \frac{\text{lat}_{\text{total}}}{\text{dep}_{\text{total}}} \right| \\
&= \left| \frac{-20 \text{ ft}}{210 \text{ ft}} \right| \\
&= 0.095238
\end{aligned}$$

Therefore,

$$\begin{aligned}
\theta &= \tan^{-1} 0.095238 \\
&= 5°26'25''
\end{aligned}$$

Azimuth α can now be found.

$$\begin{aligned}
\text{Az}_{\text{AD}} = \alpha &= 90 + \theta \\
&= 90° + 5°26'25'' \\
&= 95°26'25''
\end{aligned}$$

99 *The answer is (A).*

Find distance AB on the state-plane grid from

$$\begin{aligned}
D_{\text{grid}} &= D_{\text{measured}}(\text{sea-level scale factor}) \\
&\quad \times (\text{grid scale factor}) \\
&= (8456.9 \text{ ft})(0.999351)(0.999967) \\
&= 8451.1 \text{ ft}
\end{aligned}$$

100 *The answer is (A).*

Since MP is due west, the y coordinate of pont P is

$$y_{\text{P}} = y_{\text{M}} = 638.78$$

LN is due north; therefore,

$$\begin{aligned}
D_{\text{LN}} &= y_{\text{M}} - y_{\text{L}} \\
&= 638.78 - 540.55 \\
&= 98.23
\end{aligned}$$

Since bearing LP is N 68°35'00'' W, α is 68°35'00''.

$$\begin{aligned}
D_{\text{PN}} &= \text{LN} \tan \alpha = 98.23 \tan 68°35'00'' \\
&= 250.44
\end{aligned}$$

The x coordinate of point P is then

$$\begin{aligned}
x_{\text{P}} &= x_{\text{L}} - D_{\text{PN}} \\
&= 879.01 - 250.44 \\
&= 628.57
\end{aligned}$$

101 *The answer is (A).*

On a mass diagram, a descending curve indicates embankment and a rising curve indicates excavation.

102 *The answer is (B).*

The state-plane system is based on a cartesian coordinate system. The distortion introduced by mapping a curved surface (the earth) onto a plane would be too large if the entire state were placed on a single zone.

Office Procedures

103 *The answer is (B).*

A section measures 1 mi by 1 mi and there are four quarter sections to a section. Therefore the length of a quarter section side is a half mile. Since there are 80 chains to a mile, a quarter section side will be 40 chains in length.

104 *The answer is (A).*

The monuments on the ground and the field book records are the result of the intent of the surveyor. Neither of these can overrule the intent of the parties to a conveyance. This intent is usually expressed in the actual wording of the legal description.

105 *The answer is (B).*

The NAD adjustments were horizontal and did not affect the leveling networks. The NAVD 88 adjustment only held one point fixed—Father Point/Rimouski. NGVD 29 was adjusted to fit to 26 mean sea-level stations.

106 *The answer is (A).*

A benchmark is only associated with elevation and is defined as any point with a known elevation. A mean sea-level tidal point is a benchmark, but it does not fit a general definition.

107 *The answer is (B).*

On a vertical cliff, all contours have the same planimetric position on a map. The section AB, therefore, represents a vertical face or a vertical cliff.

108 *The answer is (B).*

Since the photo scale is 1 in:300 ft, convert the 8.5 mm to inches.

$$d_{\text{tank,in}} = (8.5 \text{ mm}) \left(\frac{1 \text{ in}}{25.4 \text{ mm}} \right)$$
$$= 0.335 \text{ in}$$

It can now be determined that 0.335 in on the photo represents on the ground

$$(0.335 \text{ in}) \left(300 \, \frac{\text{ft}}{\text{in}} \right) = 100.5 \text{ ft} \quad (100 \text{ ft})$$

109 *The answer is (C).*

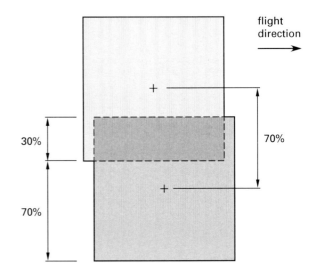

+ = photo center

The aerial photo is 9 in square. The distance between flight lines at the photo scale is

$$D = (1 - 0.3)(9 \text{ in})$$
$$= 6.3 \text{ in}$$

If 1 in at photo scale is 400 ft on the ground,

$$\text{flight line separation} = (6.3 \text{ in}) \left(400 \, \frac{\text{ft}}{\text{in}} \right)$$
$$= 2520 \text{ ft}$$

Since the scale of a USGS quad sheet is 1 in:2000 ft,

$$\begin{array}{l} \text{flight line ground} \\ \text{separation at map scale} \end{array} = (2520 \text{ ft}) \left(\frac{1 \text{ in}}{2000 \text{ ft}} \right)$$
$$= 1.26 \text{ in}$$

110 *The answer is (C).*

A quadrangle consists of an array of 4 by 4 townships, and a township is an array of 6 by 6 sections. A section is 1 mi by 1 mi, so a township is therefore 6 mi by 6 mi. It follows that a quadrangle has sides 24 mi by 24 mi.

111 *The answer is (B).*

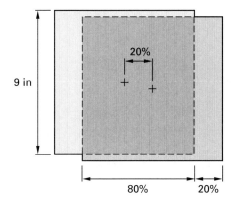

+ = photo center

The map scale is reduced 8x, from 1 in:100 ft to 1 in:800 ft, to give the photo scale. Since the photos are 9 in square, the overlap is

$$\text{overlap} = (0.8)(9 \text{ in}) = 7.2 \text{ in}$$

The distance between the photo centers is

$$D = (1 - 0.8)(9 \text{ in}) = 1.8 \text{ in}$$

1 in on the photo is 800 ft on the ground. Therefore, at ground scale,

$$1.8 \text{ in on the photo} = (1.8 \text{ in}) \left(800 \, \frac{\text{ft}}{\text{in}} \right)$$
$$= 1440 \text{ ft}$$

112 *The answer is (A).*

This definition is commonly used by surveyors to describe a metes and bounds legal description. The word "metes" refers to measurements and the word "bounds" refers to land boundaries.

113 *The answer is (B).*

Full control requires sufficient control points to level and scale the stereomodel so that a map can be compiled at an exact scale with contours. Such a stereomodel will have a minimum of three elevation points—one in each corner and two plan points with known position.

114 *The answer is (C).*

A photo image that is a true map is known as a digital orthophoto. Usually, every pixel also has an elevation above sea level. A digital orthophoto can be compiled from aerial photos or satellite images.

115 *The answer is (D).*

NAD 27 is a horizontal datum (not a vertical datum). CAN 91 does not exist. NGVD 29 is an older vertical datum. NAVD 88 is the datum described in the problem statement.

116 *The answer is (B).*

Photo scale is a ratio of camera focal length, f, to the flying height above the terrain, H.

$$H = \text{flying height above sea level} - \text{elev}_{\text{terrain}}$$
$$= 11,000 \text{ ft} - 5000 \text{ ft}$$
$$= 6000 \text{ ft}$$

Therefore,

$$\text{photo scale} = f{:}H$$
$$= 6 \text{ in}{:}6000 \text{ ft}$$

The 6 in:6000 ft photo scale can be reduced for a final answer of 1 in:1000 ft.

117 *The answer is (B).*

The method of contours directly drawn in stereo from the overlap is rarely used today, although it was the standard method in the 1950s and 1960s. Computer interpolation from stereo profiles and automated contouring by image matching are sometimes used today, but the standard method is to generate contours by computer interpolation based on a digital elevation model manually compiled on a stereoplotter.

118 *The answer is (C).*

The station of the beginning of vertical curve is

$$\text{sta}_{\text{BVC}} = \text{sta}_{\text{PVI}} - \frac{L}{2}$$
$$= 17.56 \text{ sta} - \frac{8 \text{ sta}}{2}$$
$$= 13.56 \text{ sta}$$

Find the rate of change of grade.

$$r = \frac{g_2 - g_1}{L}$$
$$= \frac{3\% - (-2\%)}{8 \text{ sta}}$$
$$= 0.625\%/\text{sta}$$

The x coordinate of the lowest point on the curve is

$$x \text{ coordinate of lowest point} = -\frac{g_1}{r}$$
$$= -\frac{-2\%}{0.625 \dfrac{\%}{\text{sta}}}$$
$$= 3.2 \text{ sta}$$

So the station of the lowest point is

$$\text{sta}_{\text{lowest point}} = \text{sta}_{\text{BVC}} + x$$
$$= \text{sta } 13{+}56.00 + \text{sta } 3{+}20.00$$
$$= \text{sta } 16{+}76.00$$

119 *The answer is (A).*

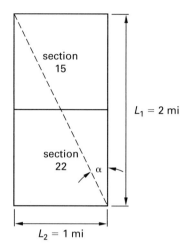

The length of the side of each section is 1 mi. The required bearing is N α W.

$$\tan \alpha = \frac{L_1}{L_2} = \frac{1 \text{ mi}}{2 \text{ mi}}$$
$$= 0.5$$

Rearranging,

$$\alpha = \tan^{-1} 0.5$$
$$= 26°34'$$

It follows that the bearing N α W is N 26°34′ W.

Note that options (C) and (D) are not bearings, since all bearing start at N or S.

120 *The answer is (B).*

It is too expensive to survey ground control points on every aerial photo for the purpose of orientation of the stereo overlaps for photogrammetric mapping. Surveyed control points are usually only placed on every second or third overlap. This means that artificial points (small drill holes) are drilled into the emulsion surface on overlaps that do not have sufficient control for mapping. These artificial points are digitized and go through a complex series of transformations, and the total process is known as aerial triangualtion.